On Fracking

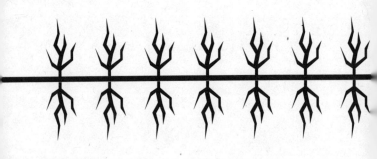

ON FRACKING

C. Alexia Lane

Copyright © 2013 C. Alexia Lane
All rights reserved. No part of this publication may be reproduced, stored in a retrieval system, or transmitted in any form or by any means – electronic, mechanical, audio recording, or otherwise – without the written permission of the publisher or a photocopying licence from Access Copyright, Toronto, Canada.

Rocky Mountain Books
www.rmbooks.com

Library and Archives Canada Cataloguing in Publication

Lane, C. Alexia, author
    On fracking / C. Alexia Lane.

(RMB manifesto series)
Includes bibliographical references.
Issued in print and electronic formats.
ISBN 978-1-927330-80-7 (bound).— ISBN 978-1-927330-81-4 (html).—
ISBN 978-1-927330-82-1 (pdf)

    1. Hydraulic fracturing—Environmental aspects.
I. Title. II. Series: RMB manifesto series

TD195.G3L35 2013   333.8'2314   C2013-903117-0
                                             C2013-903118-9

Printed in Canada

Rocky Mountain Books acknowledges the financial support for its publishing program from the Government of Canada through the Canada Book Fund (CBF) and the Canada Council for the Arts, and from the province of British Columbia through the British Columbia Arts Council and the Book Publishing Tax Credit.

 Canadian Heritage / Patrimoine canadien   Canada Council for the Arts / Conseil des Arts du Canada

 BRITISH COLUMBIA ARTS COUNCIL
Supported by the Province of British Columbia

The interior pages of this book have been produced on 100% post-consumer recycled paper, processed chlorine free and printed with vegetable-based dyes.

 MIX
Paper from responsible sources
FSC® C016245

Contents

List of Abbreviations	VII
Acknowledgements	IX
Introduction	1
That Was Then ...	7
This Is Now ...	27
The Wild Wests: Governance of Water and Energy in North America	43
Environmental and Health Concerns	81
Groundwater	91
Future Focus	99
Endnotes	107
Bibliography	115

List of Abbreviations

CAPP Canadian Association of Petroleum Producers

CEPA Canadian Environmental Protection Act, 1999

CWA Clean Water Act (U.S.)

DE diatomaceous earth

EC Environment Canada

ERCB Energy Resources Conservation Board (Alberta)

FITFIR first in time, first in right

FRAC Fracturing Responsibility and Awareness of Chemicals [Act of 2011] (U.S.)

GCD groundwater conservation district (Texas)

GHG greenhouse gas

GLWQA Great Lakes Water Quality Agreement

GWPC	Ground Water Protection Council (U.S.)
IJC	International Joint Commission
IOGCC	Interstate Oil & Gas Compact Commission
MCL	maximum contaminant level
MCLG	maximum contaminant level goal
MEOR	microbial enhanced oil recovery
MPWA	Mighty Peace Watershed Alliance
NRTA	Natural Resources Transfer Agreements
OPEC	Organization of the Petroleum Exporting Countries
PFRA	Prairie Farm Rehabilitation Administration
PNGCB	Petroleum and Natural Gas Conservation Board
RRC	Railroad Commission of Texas
SDWA	Safe Drinking Water Act (U.S.)
UIC	underground injection control
USEPA	United States Environmental Protection Agency
WPAC	Watershed Planning & Advisory Council

Acknowledgements

I wish to express my sincerest appreciation to Bob Sandford for his confidence, support and trust in me. Additionally, I would like to offer special thanks to Isabelle Netto, Robert Ferrari, Donna Sinnett, Andrew Sousa and Nick for their eyes and ideas. And to the Baymaks, Ferraris, Joneses, Lanes and Rafas for their support in untold ways.

Introduction

Hundreds of millions of years ago, hydrocarbon deposits began forming from sediments rich in terrestrial and marine organic matter. Over time, tectonic activity buried these materials beneath the Earth's surface in sedimentary rock layers, where they were exposed to pressure and heat, causing the formation of solid (coal), liquid (oil) or gaseous (natural gas) fossil fuels.

Coinciding with the formation of these subsurface hydrocarbon deposits, water-containing basins and aquifers were also formed. And not only is this connection between fossil fuels and surface and ground water a geological one that dates back millions of years, it is also present in our own brief history as a country. The earliest documentation of fossil fuels in Canada is from the North, where water-borne explorers such as Alexander Mackenzie and James Knight observed the natives using bitumen from the riverbanks to patch their canoes and for medicinal purposes.[1] Similarly, in the southern U.S., oil was used to combat rheumatic pains and as a topical treatment for sores and burns.[2] However, it was not until a century later that we entered the "Golden Era," understood to be the modern-day Age of Oil.

Hydrocarbons vary in density and physical properties, which affects how easy or difficult they may be to extract from the various deposits in which they are found. Due to their fluidity, oil and natural gas can move through the various subsurface sedimentary layers until they reach impermeable layers, underground reservoirs or the surface. One source rock for many fossil fuel deposits being developed today is shale, a common sedimentary rock which possesses relatively high amounts of organic matter. However, shale is also relatively impermeable, which means that fossil fuels trapped within its pores are not easy to extract.

Traditionally, conventional hydrocarbon deposits – crude oil, natural gas, and coal – have involved extraction from shallower layers and reservoirs and from the surface. As conventional oil and gas sources dwindle while demand for fossil fuels increases, however, there has been a corresponding shift to extracting hydrocarbons from deeper, previously inaccessible unconventional deposits. Unconventional hydrocarbons include "tight" oil and gas closely bound in the pores of carbonates and sands; shale oil and gas found in the pores of shale; coal bed methane (CBM), which is natural gas residing in coal seams; and bitumen (oil associated with sand particles). Facilitating the shift from conventional to unconventional fossil fuel extraction are "new" technologies in horizontal or directional drilling and fracturing of the source rock.

Hydraulic fracturing, or fracking, has arguably ushered in a new era of non-renewable fossil fuel extraction, beginning in North America and expanding to petroleum-producing regions across the globe. Fracking is being carried out in traditional oil and gas regions and is being proposed in "new" states and provinces not previously associated with fossil fuels. The use of multi-stage fracking in conjunction with horizontal drilling to extract unconventional hydrocarbon sources has opened up previously cost-prohibitive oil and natural gas reserves. Because of these massive unconventional energy discoveries, fracking is being touted as the energy game-changer and the most promising method to ensure a "clean energy" source for the future. Fracking has allowed our zest for fossil fuels to be sustained at a time when we need to be actively pursuing non-hydrocarbon energy sources.

Fracking is a highly controversial and extremely emotive topic for concerned citizens, landowners and environmental groups alike. Of primary concern is the protection of freshwater sources that are being used and/or drilled through in fracking operations. Current practices require the injection of large volumes of fresh water, and many wells pass through or are situated relatively close to groundwater aquifers, i.e., drinking water sources. This controversial practice has sparked an emotionally charged outcry from groups seeking legal assurance that surface water and groundwater sources are being safeguarded. There is a prevailing lack of public confidence in the regulatory structure and political bodies charged with protecting our water sources, which in part is attributed to the known extent and intensity with which wells are being indiscriminately fracked across the continent. What's more are the inextricable ties between water and energy. Energy is required to fuel our drinking water distribution networks from source to tap, and water is the critical component in extracting unconventional fossil fuel sources. Furthermore, drinking water distribution networks are at times employed in supplying water to fracking operations – this at a time when our drinking water infrastructure is crumbling, in places nearing complete deterioration.

Of the many environmental impacts and public health issues that have been raised in both scientific and "grey" literature and in public consultations about fracking, some are outside our scope here, such as noise and air pollution and localized seismic activity. This manifesto will look at the potential contamination of groundwater sources; the potential ecotoxicological effects from fracking, climate change and water management initiatives; and the bioengineering of organisms to enhance shale gas extraction.

In order to understand how the extraction of unconventional fossil fuel sources has emerged as the continental energy game-changer, though, we first need to examine the history of the oil and gas industry in North America.

That Was Then ...

The North American oil era began in the mid-nineteenth century with the first intentionally drilled well at Titusville, Pennsylvania. Colonel Edwin L. Drake's successful discovery and capture of oil heralded James Miller Williams's crude oil discovery at Oil Springs, Ontario, soon thereafter. Prior to these discoveries, oil and gas were captured where they naturally flowed to the surface or were inadvertently found when drilling wells for water or, more frequently, salt. From their modest origins in salt brine drilling, morphing into petroleum discoveries, the oil and gas industries of Canada and the U.S. have developed in tandem ever since.

Some argue that what made Drake's well so successful was the newfound ability to drill, penetrating through rock, which enabled the extraction of oil from much greater depths. Others argue that the success at Titusville can be attributed to the surge of investment in oil and oil-related products, which supported rapid industrialization. With these discoveries and burgeoning industries in the midwestern states and southern Ontario coinciding with an increased demand for more and better lighting in expanding cities, early oil ventures focused on producing kerosene. During the refining process, gasoline was formed as a by-product and often discarded, an interesting fate given the reliance we have on gasoline today. As North American economies grew and prospered, oil was established as a "key strategic commodity." In part, this fame can be attributed to its increased use in transportation sectors: naval fleets being converted from coal- to oil-powered vessels, the rise in domestic automobile purchases, the mechanization of agriculture and the conversion of railway engines from coal-fired steam to diesel.

With these new drilling technologies and increased demand for petroleum products in Canada and the U.S., very large businesses began to grow. Standard Oil (Cleveland, Ohio), which was to become the monopoly in oil and the first corporate trust, was founded by John D. Rockefeller in 1870. Not many years later, Standard Oil was out-competing its rivals, using underhanded, dubious, even violent tactics to dominate the U.S. oil industry and establish its stronghold in the global energy market. At its peak, Standard Oil is said to have controlled 90 per cent of all oil produced in the U.S. and 67 per cent of the world's oil supply. The amalgamation of Standard Oil and its affiliates into the Standard Oil Trust in 1882 ultimately led to its demise in 1911, when after years of litigation the conglomerate was found to be in violation of the federal Sherman Antitrust Act. The ruling resulted in dissolution of the trust, whereupon 34 independent companies were formed, including predecessors of present-day major players ExxonMobil and Chevron. Despite the dissolution of the trust, the business model it embodied provides a classic example of the economic and innovation ramifications of monopolies. The Standard Oil empire grew undeterred for decades because of its use of illegal and corrupt methods to gain industry advantage and out-compete other companies while stifling innovation. We see these tactics carried forward into monopolies today.

In response to Rockefeller's Standard Oil,

several Ontario oil producing and refining companies merged to form Imperial Oil. At the time, Canada was relying heavily upon U.S. exports to supplement its dwindling oil supplies. In alignment with Standard Oil's ever-widening empire, Imperial was acquired by a Standard Oil holding company, further strengthening Standard's position in the global refining arena.

Both Standard and Imperial understood the importance of water to the success of their endeavours: Standard established itself in Cleveland, Ohio, on Lake Erie, while Imperial set up at the southern tip of Lake Huron. Water is required in the refining process, and being located on the Great Lakes also facilitated the transportation of oil and associated products. Moreover, in the case of Imperial, much-needed oil supplies from the Midwest were readily accessible through the Great Lakes network to ensure that adequate quantities were available to Canadians.

For several reasons, 1911 was a monumental year in the oil era: the Standard Oil trust was dissolved, gasoline surpassed kerosene as the dominant product from oil refining, and the British Royal Navy converted its fleet from coal to oil. The influence of the navy's decision, together with the diminishing of oil supplies in Ontario, prompted the Canadian federal government to grant incentives to the oil industry to seek out and develop domestic reserves. Thus the Canadian industry moved west and north. The U.S. had already discovered large petroleum reserves in its western regions; extraction was well underway in Texas, Oklahoma, Kansas and Louisiana.

As the petroleum industry blossomed, waste management practices were born out of the prevailing parochial worldview of the time, the dilution paradigm: dilution is the solution to pollution. The dilution paradigm appropriately aligned with petroleum management activities, with waste often buried in a pit on a lease site or discarded into a nearby water body without discretion. These environmentally destructive practices continued well into the 1970s, until the establishment of overarching federal environmental regulatory bodies.

The U.S.'s Gusher Age, or Texas Oil Boom, from 1895 to 1945, was kick-started by the Lucas gusher in 1901 at Spindletop, near Beaumont, Texas, on the Gulf coast, initiating "Uncle Sam's oil fame" as one reporter put it. Spindletop launched the oil and gas industry in Texas, which remains among the most prolific producing states to this day, not to mention being, along with Kansas and Oklahoma, the birthplace of modern fracking. Meanwhile, in Canada, discoveries were made at Norman Wells, Northwest Territories, and Turner Valley, Alberta. However, it was not until the massive Leduc discovery in Alberta in 1947 that Canada was able to move away from its reliance on imported crude oil from the U.S., a phenomenon that continues today. Just as Spindletop ushered in a new era of petroleum development in Texas, so too did the Leduc discovery in Alberta. The parallels between Texas and Alberta were already becoming apparent.

Prior to their respective booms, both jurisdictions had agriculture-based economies that relied on irrigation; when oil was struck, they rapidly transitioned to an economic insulation that continues today. Although Alberta lagged behind Texas by nearly half a century in striking oil, the unconventional resource revolution has considerably closed the gap between them. Texas set the precedent for the development of shale gas with the Barnett play and has continued with the Eagle Ford and Haynesville plays, whereas Alberta has focused on extracting bitumen from its northern oil sands and is on the cusp of expanding development in some of its larger unconventional plays. As both Texas and Alberta move forward, albeit in different pursuits, their connection is further strengthened by the proposed Keystone XL pipeline. Both state and province have reaped untold financial benefits from their respective hydrocarbon sectors.

There are multiple examples of states, provinces and territories, including Texas and Alberta, in which petroleum-based economies have trumped traditional agricultural economies. Often these agricultural industries require irrigation in order to produce viable crops, be it from surface water or groundwater. To date, water has been predominantly allocated from groundwater in Texas and surface water in Alberta; however, we know that unconventional resource development and changing hydrology across the continent are going to require a reassessment of water management practices. In order to proactively ensure adequate freshwater resources for drinking and food production, a temporary hiatus in unconventional resource development – until new predictive models for changing hydrology and weather patterns can be developed – would go a long way toward securing these integral water sources for generations to come. Scaling back unconventional development today would also allow for innovation of new technologies and techniques to coax unconventional resources that are less water-intensive and therefore sustainable.

With exploration came innovation. With oil extraction well under way in both Canada and the U.S., drilling and removal methods were revised to maximize extraction and exhaust conventional reservoirs; unconventional extraction was a secondary priority. The first recorded use of fracking (vertical process) was made by Stanolind, a predecessor of Amoco, in the late 1940s in Kansas, soon followed by Halliburton in Oklahoma and Texas. Accelerated fossil fuel transportation began in the 1950s, when extensive pipelines to the coasts and across the continent were constructed, enabling oil and natural gas to be disseminated to areas previously inaccessible due to lack of transportation infrastructure. Many of these original lines are still in use, with miles more pipe being added annually. The extent of these fossil fuel pipelines preceded the expansive drinking water distribution infrastructure – pipelines – that now exist across North America. Certainly, fossil fuel lines cover longer distances and extend farther into isolated, sparsely populated areas, where drinking water is often derived from an on-site well or cistern. Ponder that fact: oil and gas are more readily accessible than drinking water!

The 1950s and 1960s saw undeterred petroleum extraction across the continent. Several key events that occurred during these two decades had far-reaching effects: the formation of the intergovernmental Organization of the Petroleum Exporting Countries (OPEC) in the Middle East, M. King Hubbert's hypothesis of peak oil, and the 1962 publication of Rachel Carson's landmark book *Silent Spring*. The ramifications of these events are perpetuated in present-day energy market trends, practices and environmental ideals.

OPEC

OPEC grew during the 1960s but garnered global attention throughout the 1970s when its member countries took control of their domestic petroleum supplies, causing global oil prices to soar and creating international market tumult. This shift in power balance stemmed from several geopolitical events in the Middle East: the Arab oil embargo, the Iranian revolution and the lead-up to the Iran–Iraq war. Owing to its everyday importance, oil had previously established itself as a globally priced commodity. With events in the Middle East causing fluctuations in North American markets, it quickly became apparent that securing domestic petroleum supplies was of paramount importance to North American autonomy. Surprisingly, given the known endowment of fossil fuels across the continent, there were barriers to ensuring an insulated continental fossil fuel supply: in Canada the National Energy Program (1981) was developed to guarantee that fossil fuels extracted in the west (primarily Alberta) were prioritized to the eastern provinces at less than a market-value price; and in the U.S., conventional peak oil was attained at the start of the 1970s, instigating closer ties with Canada to secure future fossil fuel supplies from an ally. No doubt this neighbourly gesture will be repeated as water resources in the U.S. decline more rapidly and states begin turning north in an attempt to secure water from Canada. It is only a matter of time before the discussion around the commoditization of water is reinvigorated in the international arena, given the known decrease in water supplies around the world.

What may or may not have been projected during this time of tumult was the ever-increasing demand for fossil fuels. However, we need only look around us to see that there has been an exponential rise in demand for fossil fuels during the last 40 years, and it continues to increase. With the global population having surpassed seven billion and with the rapid industrialization of developing economies, including China, India and Brazil, it no longer serves us to reside in a place of denial – our appetites for fossil fuels are insatiable. The current unconventional resource revolution only serves to bolster a false sense of security that our petroleum and water resources will somehow remain plentiful and intact.

Peak oil

The American geophysicist M. King Hubbert logically hypothesized that because oil – or any fossil fuel, for that matter – is a non-renewable natural resource, its development would follow a trajectory resembling the typical bell-shaped curve, beginning with discovery, peaking at maximum extraction and subsiding to zero with exhaustion of the resource years later.[3] The concept of the end of oil and images of a post-apocalyptic world have often been portrayed in popular media to effectively induce fear; however, this has not prompted industry to seek out alternatives to fossil fuels in any tangible way as one would have hoped. Instead industry has pushed forward, developing petroleum reserves at unprecedented rates for decades. This industry push coincides with the prevailing societal belief that fossil fuel resources are infinite. It is important to acknowledge that the same societal belief can be applied to our view of water as an infinitely abundant resource. While industry fights to keep up with societal demands, the pleas by environmentalists to slow development to a sustainable pace are often overlooked in favour of immediate economic growth. For too many decades this has been the norm in North America. Now that we have entered a new age of energy extraction that further threatens our natural environment, what legal recourse is there to moderate the pace of industry development? As previously mentioned, the U.S. reached its peak production of conventional oil in the early 1970s, and the global conventional peak was reached in 2006. Given that we are now on the decline from peak oil, and in some places have been for decades, there was a natural progression toward seeking out alternative energy sources. However, this does not mean that alternatives to non-renewable fossil fuels were explored in depth. In North America it has meant further development and modification of technologies in order to extract previously unreachable fossil fuels, including but not limited to shale reserves and oil sands.

In the shadow of peak oil, the consideration of alternative energy sources (i.e., renewables) is becoming increasingly important. To clarify, renewable energy sources or alternatives to fossil fuels must be independent of hydrocarbons for their production; this diminishes the integrity of wind and solar as renewables. In fact, within these parameters of non-dependence on hydrocarbons, biomass has demonstrated the greatest potential as a future renewable energy supply.[4] But years of research and testing will be required in order to fully understand biomass's potential and maximize its efficiency. Now is the time to allocate financial and scientific resources to understanding alternative energy options. Instead, federal funding for scientific research is being cut and millions of dollars are spent on each fracked well to continue pursuing non-renewable energy.

The environmental movement

In the wake of multiple global environmental disasters such as atomic-bomb testing, smog, DDT dumping and others, Rachel Carson's peer-reviewed book *Silent Spring* (1962), together with grassroots initiatives, pioneered the environmental movement as we know it today. Carson raised awareness of the effects of several recalcitrant pesticides and chemicals on ecosystems and health. Interestingly, Carson detailed the adverse effects of some toxic and carcinogenic chemicals (e.g., benzene) that are known to be used in large quantities in fracking operations today. Considering that this information was documented and made publicly available in *Silent Spring* and other literature more than fifty years ago, it is difficult to understand how the continued use of these chemicals is condoned. We know the adverse impact on our ecosystems, and that this in turn has an adverse impact on us, and yet there is no legal deterrent in place to halt the use of these chemicals in fracking operations.

Accompanying the growing public concern for the environment was a shift away from the dilution paradigm ("dilution is the solution to pollution") and toward the boomerang paradigm: what you throw away can come back and hurt you. With the boomerang paradigm fresh in the public mind, the impetus to establish an enforceable environmental regulatory structure was born. In both the U.S. and Canada, federal environmental bodies were created: the United States Environmental Protection Agency (USEPA) in 1970 and Environment Canada (EC) in 1971. These overseeing bodies remain active today, drafting environmental policy and guidelines and enforcing the relevant statutes and regulations with the goal of protecting the natural environment.

Because fossil fuel development affects several aspects of the natural environment, the governance of industry activities is shared among federal, state, provincial/territorial and/or municipal jurisdictions, resulting in a convoluted political and regulatory structure that can be difficult to understand and unravel. Accountability for specific acts with far-reaching environmental impacts is difficult to apply to any one body or jurisdiction, further complicating enforceability or penalization. Traditionally, this power split resulted in delays to the mandated approval process, but there is now a movement toward streamlined, "single-window" regulatory procedures to expedite resource extraction. The one-window approach is effective in supporting industry activities, but unfortunately it makes environmental and public health concerns a secondary consideration in decision-making. This new approach only serves to heighten public distrust of regulatory bodies charged with protecting public and environmental health.

Stemming from soaring oil prices in the 1970s and a worldwide recession in the early 1980s, petroleum production and consumption declined globally during the early 1980s. Markets were starting to recover and production was beginning to turn upward when the 1986 oil market crash occurred. Because of political pressure and jostling, Saudi Arabia (head of OPEC) flooded the market, causing global prices to tumble. Despite the upheaval caused by OPEC in the 1970s, the 1986 market crash shocked the North American economies, resulting in losses of billions of dollars in oil revenues and hundreds of thousands of jobs. Eventually, North American markets recovered and prices stabilized through the 1990s, with a simultaneous increase in fossil fuel demand.

A turning point in Canadian natural gas history occurred in 1986. Deregulation of the gas industry, in conjunction with broadening North American free trade, marked the beginning of increased natural gas production and export to the U.S. Pipeline systems were greatly expanded to transport increasing volumes of natural gas from Canada to the U.S. Canadian gas production doubled from 1986 to 2001, while exports to the U.S. quadrupled during the same time.[5] It is important to note that relatively cheap fuel prices during this time contributed to the societal view of fossil fuel abundance – reserves were touted as endless. While Canadians were busy producing and exporting natural gas to the U.S., American companies were experimenting with extraction techniques to unlock known shale gas reserves, beginning with the Barnett play in Texas. As we now know, successful experimentation and the application of modified drilling techniques led to the unconventional resource revolution currently sweeping across North America.

We also know we are in the midst of a global decline in conventional fossil fuel extraction and have been for decades. Nevertheless, increasingly limited conventional development continues, with known reserves being depleted too quickly to satisfy the demand. At this juncture it is important to note the parallel between conventional fossil fuels and global groundwater aquifers, which are being drained faster than they can possibly recharge. As a result of declining conventional fossil fuel sources, the development of unconventional reserves, such as Alberta's oil sands and the U.S.'s oil shales are well underway. Shale gas extraction is setting an unprecedented pace of development. Owing to industry foresight and economic impetus, we have seamlessly transitioned from declining conventional fossil fuel extraction to full-blown unconventional gas extraction – CBM, tight and shale gases – with little to no limits upon the intensity and pace at which these unconventional sources are extracted.

This Is Now ...

The highly controversial method called fracking (multi-stage hydraulic fracturing of underground rock strata in conjunction with directional drilling – downward, then horizontally or at an angle into the desired formation) is being employed to access "new" or unconventional natural gas plays. Wells are often fracked several times. Fracking as a means of extracting natural gas from shale plays (shale gas) is setting new precedents. The technique has literally cracked natural gas extraction wide open.

Quite simply, fracking is the high-pressure injection of large volumes of water (usually fresh water) mixed with chemical additives, biocide and a granular "proppant" such as sand or small ceramic beads to keep the resulting rock fissures propped open to release the resource. Ultimately the rock's natural pressure is exceeded by this injected "cocktail," "fracturing" the rock to release the fossil fuel so it can be collected and pumped to the surface via the well bore. Once the frack is complete, the injected fluid – now called flowback water (the liquid, including water, that returns to the surface relatively quickly after pressure is removed) and produced water (fluid that returns to the surface throughout the life of the well) – is guided through the well casing to the surface, where it is then stored; recycled or disposed of by underground injection; discharged into surface water; or treated.[1] It can take months for injected water to return to the surface, and estimates of the amount that does come back range from 15 per cent to 80 per cent,[2] a broad variance. Due to the combination of horizontal drilling and hydraulic fracturing, the area affected by a single well and its radiating fractures can be several square kilometres. Just as aquifers occur at varying depths below ground, so too do shale plays. It can be assumed that the deeper the play, the more taxing the extraction will be, both financially and on resources. Thicker cement well-casings, more water, more chemicals and more time will be required to get at deeper shale gas reserves. Several of the shallower plays have essentially been exhausted over the past twenty years, and unconventional extraction must continue drilling deeper and deeper below the Earth's surface.

The U.S. is setting the global pace for shale gas extraction because of advances in fracking technology over the past two decades. Fracking is underway in more than half of the lower 48 states and in Alaska. It is gaining momentum in Canada, Europe, Asia and South America. Not surprisingly, Texas and Alberta are leading the way in fracking intensity and extent. The Barnett play, which partly underlies the greater Dallas–Fort Worth area, heralded the boom of unconventional resource extraction, with intensive drilling beginning in the late 1990s.

While oil is a globally priced commodity, the value of natural gas is set by localized market demand. Gas from shale plays is being touted as America's "clean energy" source and promoted as assuring future energy security. The labelling of natural gas as "clean energy" is deceptive, however. While natural gas does have less carbon intensity than oil or coal, it is far from clean or carbon-neutral. And the amount of energy it takes to produce fossil fuels from unconventional plays, along with the consequent greenhouse gas (GHG) emissions, often greatly exceeds the energy required for extracting conventional hydrocarbons. What's more, because the shale gas revolution is now producing more natural gas than the market can burn or economically liquefy and export, prices in North America are at record lows. Obviously, this does very little to promote conservation or sustainable resource development. Rather, it favours immediate industry profit under the guise of job creation, as opposed to longer-term environmental preservation.

For the most part, the U.S. has been importing oil since the 1970s, to ensure adequate supply for ever-increasing demand, so it is understandable from an energy security perspective that we are seeing a burgeoning of natural gas extraction made possible by fracking. However, the pursuit of this "secure energy" still hinges on a non-renewable natural resource! Large volumes of finite fresh water are exploited to extract this dwindling supply of fossil fuels while jeopardizing water – both surface and ground – the only compound we cannot live without. In this vein, it is difficult to sit back and watch the energy industry and regulatory bodies open Pandora's box. Moreover, energy industry analysts understandably noted for bullish market forecasts are in some cases retracting their optimistic projections as information pours in on fracked wells' lifespans and the energy requirements of extracting from unconventional sources. Since it has been predicted that the U.S. will become a net exporter of natural gas by 2021 because of unconventional extraction,[3] it is important to question where this seemingly unlimited supply of non-renewable natural gas is to be derived from. Are these predictions based on extrapolated data of past use, or are they based on current use and projected demand? Two very different scenarios. The notion of an energy-independent U.S. is difficult to grapple with, given the ever-increasing demand for fossil fuels and the known decline in gas production and water resources across several states, coinciding with weak energy conservation initiatives (at best). Where is all the water required to extract these unconventional energy sources going to come from, and who is going to relinquish their water rights to known competing interests? We need only remember the 2011 drought in Texas, when bidding wars were carried out between irrigators and oil and gas developers to secure diminished water resources, to remind ourselves of how quickly water becomes marketed as a commodity to competing interests, despite being a "commodity" that has no possible replacement.[4]

Fracking threatens the integrity of both our surface water and our groundwater sources. Water injected into the well may be acquired from surface water such as lakes, rivers, ponds or wetlands; groundwater sources; private wells; and/or municipal water supplies. Vertically and horizontally drilled wells often traverse aquifers to reach the shale formation or they may be situated near aquifers. Fracking practices to date have relied heavily on the quality of the engineering and construction of the well casings to be adequate to prevent leakage of contaminants into groundwater. The layers of cement used are often cited as insurance that no contamination of aquifers will occur. Understandably, invoking layers of cement as a means of assuaging citizen angst is not adequate; citizens and landowners are calling for more robust assurances to preserve the integrity of groundwater sources for generations to come.

When considering the relative newness of the field of horizontal fracking combined with the extent and intensity with which this is now being done at wells all across the continent, it is in our best interest to question how these well bores and associated infrastructure are designed and built. We know that inherent in a directionally drilled well bore is the angle or curve, deep beneath the surface, where drilling orientation transitions from vertical to horizontal. Undoubtedly there is an excessive amount of pressure where the transition occurs. Is this transition zone reinforced? If so, how? How can we be sure that over time these lengths of cemented casing will remain in place and intact as intended? Shifting and settling of the strata is inevitable, as is naturally occurring microbial degradation of the infrastructure. Additionally, subsurface strata are not uniform; the well bores will encounter natural fissures, holes or caves. These geologic irregularities may consume more cement than anticipated, in turn compromising the structural integrity of the entire well bore if they are not taken into account from the outset of design.

In the evolution of engineering design, a body

of knowledge accumulates from accidents and near misses, giving rise to more redundant controls and safer operating practices. Given the relatively short history of fracking and the engineering innovation that has enabled current practices, how many near misses or modes of failure have been incorporated? Have engineering controls been actively introduced to mitigate near misses? Have the causes and outcomes of these events been catalogued and shared among fracking cohorts to prevent repeat malfunctions? Has industry allocated sufficient resources to a comprehensive failure modes and effects analysis? This common method of risk analysis in complex systems would be applicable to fracking operations. Assuming that the well bores and infrastructure are engineered to be redundant, what add-on controls, if any, are in place to protect the surrounding geology, including groundwater, in the event of compromised infrastructure? It is important that these questions be addressed by an independent, reliable agency in order to bolster public confidence and avoid all-too-common accusations of inside-the-fence self-monitoring by industry.

Because there is hydrologic connectivity

between surface water and groundwater, it is important to take into account the inverse relationship created by our actions: as our surface waters become depleted, our reliance on groundwater will only increase. Therefore we must be vigilant in protecting these integral subsurface water sources. These fossil fuel resource stresses are imposed on our water sources at a time when we are being forced to re-evaluate our antiquated water management practices in favour of new strategies that align with current trends and conditions, including the effects of fracking on surface water and groundwater. At the time many water management frameworks were drafted, they were based on the principle of hydrologic stationarity, whereby "natural systems fluctuate within an unchanging envelope of variability."[5] Now we must attempt to mitigate the impacts of anthropogenic climate change, which have contributed to the loss of hydrologic stationarity – a game-changer in water management. Current water management structures were put in place well before the water-intensive unconventional-extraction revolution was launched. Regulators and policy-makers must ensure that robust measures are in place to safeguard our water resources before it is too late. With the massive volumes of surface water used in fracking, it is more important than ever to ensure that groundwater sources are protected.

To date, no undeniable, proven cases of groundwater contamination can be directly linked to fracking, despite the massive outpouring of anecdotal evidence. Highly publicized claims of groundwater contamination suspected to have been caused by local fracking operations have arisen in Texas, Wyoming, Pennsylvania, Alabama, Colorado, West Virginia and Alberta. The lack of proven cases has done very little to allay the fears of environmental groups and citizens, in part as a result of claims being discarded because of semantics in definitions of fracking rather than because of rigorous scientific evidence. There is no assurance that contamination is completely avoidable or that the necessary legal framework is in place to prevent such occurrences.

Consider the well-publicized case of drinking water contamination in Pavillion, Wyoming, where local supplies were alleged to have been contaminated by natural gas activity in the region, as supported by preliminary findings in a

USEPA investigation. Results from water quality testing of both private and municipal wells demonstrated that the water contained chemicals commonly associated with fracking. However, these findings were disputed by the Canadian-owned drilling company because there was no baseline data to definitively link local natural gas activity to groundwater (drinking water) contamination.[6] The case highlighted the weakness of the regulatory structure that purports to protect our drinking water. The take-home lesson for landowners is to be diligent in obtaining baseline groundwater information from certified laboratories. National groundwater inventories in Canada are incomplete, and in the event your groundwater becomes contaminated from fracking, this baseline data will provide a reference point against which pre- and post-contamination water quality parameters can be tested. Without this data, it is impossible to scientifically demonstrate that water quality has been compromised by fracking.

Additionally, both quantity and quality of water need to be considered when evaluating the effects of fracking on water resources. Millions of litres of fresh potable water is used in each well, with possibly half of it returning to the surface. Knowing that this fracking fluid is predominantly fresh water mixed with chemicals and a proppant, what percentage of these millions of litres of water is able to be recycled and reused in subsequent fracking operations? More importantly, what percentage of this water is regulatorily required to be reused? Can the produced and flowback waters be adequately treated and safely returned to surface water bodies? Herein lies another important consideration with regard to the chemical mixture: are there any means or technologies capable of removing, remediating or reclaiming the hundreds of chemicals proven to be used in fracking? There has been a surge in service companies that extract and resell marketable chemicals derived from the fracking-fluid cocktail, but flowback and produced waters contain many more chemicals to be dealt with. The market for wastewater treatment in North America is rapidly expanding to accommodate the flux of produced and flowback water generated in fracking operations. From an ecotoxicological standpoint, one must question the long-term and synergistic effects of these potentially harmful substances on our populations and ecosystems.

No congruent regulatory/legislative framework is in place to oversee fracking. Many of the plays being developed traverse the Canada–U.S. border or interstate or interprovincial boundaries, which complicates the process of governing extraction and ensuring the protection of water resources. The differing perspectives on extraction laws and practices between states and provinces can cause friction. For example, the expansive Marcellus play, which underlies much of Pennsylvania, New York, West Virginia and parts of Ohio, Maryland, Tennessee, Virginia and Kentucky,[7] also extends across the Canada–U.S. border to underlie parts of Quebec and Ontario. The Marcellus is being extensively drilled in all associated states and provinces to various extents, save for New York and Quebec, where the precautionary principle has driven moratoria until environmental assessments have been completed. These moratoria are rapidly being reconsidered as the environmental assessment deadlines loom and pressure mounts on both federal and neighbouring governments from companies wishing to exploit known shale gas reserves with unencumbered drilling permits. Understanding that the hydrogeology of these plays is continuous and does not adhere to man-made boundaries, how can the impact of drilling in one state or province be constrained to that locale only and not traverse into regions where drilling moratoria are in place? It cannot. Given the interconnectivity of these resources, the withdrawal of water from one location will have far-reaching impacts downstream, depending on the location of the water body within the watershed. And we know that watersheds, water bodies and groundwater traverse boundaries – the same boundaries these shale plays traverse. It is time for policy-makers, governments and regulators to collaborate and incorporate the interconnectivity of our geological and water resources into a sound legislative and regulatory framework aimed at safeguarding our overtaxed water resources.

These newly accessible unconventional reserves have led to a reassessment of Canadian and U.S. national wealth. But at what cost are unconventional sources extracted, given the excessive use of fresh water and the corresponding chemical cocktails that are required for releasing the oil, gas or

CBM from non-porous rock and seams at various depths? It is well documented that fracked wells result in a larger overall ecological footprint than their conventional counterparts. To date, however, it has usually been only industry representatives assuring the public that contamination cannot happen, which only serves to diminish the already waning public confidence in our politicians and legislative frameworks. At a bare minimum, we need tangible evidence that our laws protect us and the environment, not industry.

The Wild Wests: Governance of Water and Energy in North America

The governance of energy and water in North America is convoluted and spans federal, state, provincial, territorial, regional and municipal jurisdictions. Several U.S. and Canadian federal laws pertaining to energy and water regulation are closely associated and intertwined, and in some aspects they intersect. Water is not governed based on hydrologic connectivity, which would at least give it some continuity, but instead is overseen by piecemeal legislation and regulation that is fragmented into various sources and components, enormously complicating its management. Because water and energy development at times cross international, man-made boundaries, conflicting development plans may arise. In light of the numerous and overlapping levels of government involved, achieving consensus to amend existing laws – for example, to take into account the loss of hydrologic stationarity – is nearly impossible. And as we see today, the struggle to balance the contending imperatives of economic development and environmental conservation is being won by the former. Although it can be argued that environmental laws and practices have vastly improved in the past 150 years of petroleum development, what must be acknowledged is the magnitude of existing and emerging challenges pressing upon the environment: population growth, the pace of development, greenhouse gas (GHG) emissions linked to industrialization, and climate change, to name but a few. With this in mind, it is a challenge to untangle the laws and understand jurisdiction and accountability.

Decades before the establishment of environmental laws in Canada and the U.S., both federal governments realized the importance of collaborating to safeguard the world's largest freshwater system, the Great Lakes–St. Lawrence River system. Pursuant to the 1909 Boundary Waters Treaty, the International Joint Commission (IJC) was established as a bilateral panel, with both countries equally represented, charged with reviewing problems and deciding on issues affecting the system. Partially straddling the Canada–U.S. border, the Great Lakes–St. Lawrence system is integral to industry and populations on either side; today it supplies drinking water to more than 40 million North Americans.[1] Of the eight states and two provinces directly associated with the system, at least half possess known unconventional petroleum reserves slated to be fracked or already being fracked. Without a doubt, these activities pose new threats to the ecosystem. Just as these water bodies stretch across the international border, so too do the unconventional plays that underlie them: Marcellus, Utica and Antrim, to name but a few. To date, no overarching federal law exists in Canada or the U.S. to oversee fracking, let alone any binding, enforceable legislation to address the impacts of unconventional development in international plays.

From its origins in British colonial law, the governance of water in Canada has been difficult to navigate. Water was not specifically addressed in the original constitutional statute, the British North America Act (1867), now called the Constitution Act, 1982. That being so, the evolution of water governance is fragmented, with the federal government having powers over specific aspects, including fisheries, shipping and navigation, and over water once it crosses international or interprovincial boundaries, whereas provinces possess power over management and allocation. Allocation is becoming an increasingly important aspect of water management, given the large surface water withdrawals known to be routinely made for fracking, the changes to surface hydrology from altered climatic regimes, and the need for water to irrigate vital food crops under increasingly dry conditions and to supply expanding populations. The fact that water is diffuse, and is defined by equally diffuse watersheds, also complicates the governance of it, especially when linked to a transnational environmental issue such as the potential effects of fracking on our freshwater resources.

At the federal level in Canada, the Fisheries Act (1985) and the Canada Water Act (1985) govern fish habitat and water quality and management, respectively. The Fisheries Act applies in part to substances determined to be deleterious to fish habitat – in essence, water quality – while the

Canada Water Act deals more specifically with water resource management, including "research and the planning and implementation of programs relating to the conservation, development and utilization of water resources." The most comprehensive environmental statute that broadly applies to water in Canada is the Canadian Environmental Protection Act, 1999 (CEPA), a component of which deals with toxic substances that may enter water bodies.[2]

In 1987 a Federal Water Policy was put forward with the objective "to encourage the use of freshwater in an efficient and equitable manner consistent with the social, economic and environmental needs of present and future generations." Unfortunately this comprehensive policy was never put into effect. It outlined ambitious goals that included water pollution prevention strategies and promoted the efficient management of water, and it proposed specific steps for a broad course of action.[3] Undoubtedly, if this measure had been implemented as intended, the water issues Canadians are currently contending with would be much different. When this policy was drafted 25 years ago, it was proactive. If implemented today, and I would argue that an updated version of it should be, it would be only reactive.

Another federal proposal, the Toxic Substances Management Policy (1995), presented an aggressive plan aimed at eliminating man-made poisons from the environment and managing such substances throughout their life cycle.[4] Once again, the measure was never given effect.

It is interesting that the far-reaching issues these two federal policies addressed are still relevant today, perhaps even more so given the escalated assaults on our water resources and changing hydrology. We need to reinvigorate policies such as these in order to effectively mitigate disastrous effects on our water resources. And we need to start managing water as the finite, imperative resource that it is.

At the provincial level, water is allocated by licence. This is particularly important in the Prairie provinces – Alberta, Saskatchewan and Manitoba – as they rely on runoff from the Rocky Mountains to recharge the majority of their primary potable water source: surface water. And given that these runoffs are headwatered in the Rockies, it is Alberta that is the main source of water for Saskatchewan and Manitoba. We'll take a look at Alberta's system of prior allocation of water in a moment.

With power over water residing in both federal and provincial/territorial jurisdiction, several collaborations have formed among the relevant levels of government to ensure adequate responsibility for water. For example, the interjurisdictional Prairie Provinces Water Board was formed in 1948 with representatives from the three provinces and the federal government. In 1969 the Master Agreement on Apportionment was signed to ensure sufficient water to all three Prairie provinces. To this end, 50 per cent of all natural eastward-flowing water from Alberta must cross the border to recharge Saskatchewan water sources, and 50 per cent of the natural eastward-flowing water from Saskatchewan must cross the border to recharge water in Manitoba.[5]

This snapshot of the linkage of water resources across the three provinces demonstrates just how reliant we are on proper source and upstream water management. Impacts on Alberta's water resources affect Saskatchewan's and Manitoba's as well. Recall that these surface water volumes are being reduced at rapid rates because of changing climatic conditions and growing demand from resource development. Recall also that fracking requires millions of litres per well, which is most often obtained from surface water, and that fracking is occurring or slated to occur in all three Prairie provinces, with Alberta (where the surface water originates) leading the way. Interestingly, fracking in Saskatchewan and Manitoba is in part attributed to the large, international Bakken play, so not only are there interprovincial aspects to governance, there are international implications as well.

An important distinction between water law in Canada and the U.S. is that Canada has no overarching federal laws specific to the governance of water. For example, the integrity of Canadian drinking water is dealt with by the Guidelines for Canadian Drinking Water Quality, which mandate no enforceable actions and leave implementation/adherence to the discretion of the provinces/territories. Without a national framework on water governance, Canadians are at the mercy of a largely unregulated and unmonitored system for safe water. If the safety of our drinking water is not of primary legal importance, how can we expect to be protected from the harmful effects of fracking on our water quality and quantity?

As mentioned earlier, the U.S. does have overarching federal water laws. These are overseen and enforced by regulating bodies including the USEPA, which administers the Clean Water Act (CWA) (1972) – the "cornerstone of surface water quality protection"[6] – and the Safe Drinking Water Act (SDWA) (1974), which "ensures the quality of Americans' drinking water."[7] Although the USEPA has authority over such statutes, there are also federal/state understandings in place whereby states implement and enforce day-to-day activities as set out by the USEPA.

The CWA was born from 1972 amendments to the U.S.'s original water pollution law, the Federal Water Pollution Control Act of 1948. These amendments included the establishment of a structure to regulate the discharge of pollutants into surface waters, and the granting of authority to the USEPA to implement pollution control programs. The CWA maintained power to set water quality standards for contaminants entering surface waters and recognized the importance of anticipating impacts imposed on surface waters from non-point-source pollution. The statute employs regulatory and non-regulatory tools to reduce pollutant discharge into waterways (point-source), manage pollutant runoff (non-point-source) and finance water/wastewater treatment facilities. It is important to note that the CWA excludes groundwater and water quantity; its jurisdiction is solely over surface waters. If the ultimate discharge method for post-fracking flowback water is disposal into a surface water body or by underground injection, treatment prior to discharge is regulated by the National Pollutant Discharge Elimination System, which is authorized by the CWA.

Another relevant law stemming from amendments to the Federal Water Pollution Control Act of 1948 is the Great Lakes Critical Programs Act (1990), which addresses water quality in the Great Lakes. The Great Lakes have played an integral role in petroleum development since the onset of the modern oil era; recall the situating of Standard Oil and Imperial Oil on the international freshwater bodies of Lake Erie and Lake Huron respectively. Furthermore, the Great Lakes have strengthened cooperative ties between Canada and the U.S.; the

Great Lakes Water Quality Agreement (GLWQA) was first signed in 1972 between Prime Minister Trudeau and President Nixon, with amendments in 1978, 1987 and 2012.[8] Note that both the CWA and GLWQA were established in the same year, implying a shared understanding of the importance of clean water between the two countries. From the outset, the GLWQA included recognition of the need to foster adaptive management by stating the projected mandate to continually amend the agreement as new challenges and threats emerge in the Great Lakes. The 1978 revision contained the purpose statement "to restore and maintain the chemical, physical, and biological integrity of the waters of the Great Lakes Basin Ecosystem," which demonstrated the holistic ecosystem approach intended by the agreement. The CWA included an identical purpose statement, but with broader application to all the nation's waters.

Given that both the CWA and the Great Lakes Critical Programs Act pertain to surface water quality in the U.S., there is crossover between them. The Great Lakes Critical Programs Act put into place components of the GLWQA, resulting in changes to the CWA mandating the USEPA to determine water quality criteria and set maximum levels for several persistent toxic pollutants. The GLWQA is under the authority of the International Joint Commission, but the IJC possesses no power to impose penalties or sanctions if either country fails to comply or meet the objectives outlined in the GLWQA. Thus the efficacy of the GLWQA is confined to the actions of either nation. As we have seen, it is of paramount importance that the potential effects of fracking in the Great Lakes region be taken into account as a new, emerging threat to the freshwater ecosystem. However, although the GLWQA was updated in 2012, there is no evidence that such acknowledgement is happening; both federal governments have abdicated responsibility, once again favouring industry over environment.

From its enactment in 1974, the Safe Drinking Water Act has been charged with the monumental task of, as the USEPA website puts it, "regulating the nation's public drinking water supply," including groundwater, to ensure that clean, safe drinking water is available to Americans. It is important to highlight that public drinking water supplies are defined as those serving 25 or more people;

therefore, private/domestic wells are exempt from the SDWA, and only a limited number of states have regulations in place to oversee the drinking water quality of private wells. This is critical when fracking is carried out in rural settings, as it often is. As these private wells are outside any legal jurisdiction, the onus for ensuring safe drinking water quality rests solely on the owner, hence the importance of obtaining baseline groundwater quality information. The statute charges the USEPA with setting national drinking water standards for both naturally occurring and man-made contaminants discovered in drinking water; however, these standards are also monitored by states, tribes and the public. One must consider the magnitude of this undertaking, knowing that it concerns drinking water safety for hundreds of millions of citizens, with thousands of potential contaminants to monitor and track throughout the more than 160,000 public water systems. Amendments were made to the SDWA first in 1986 and again in 1996. The 1996 changes expanded the law to include not only treatment of water but protection of source water, operator training, infrastructure improvement funding and the inclusion of public information. Furthermore, the SDWA establishes a framework for an Underground Injection Control (UIC) program, which is charged with regulating the injection of wastes into groundwater. The linkage between the SDWA and the UIC program as it pertains to fracking has become a contentious issue.

The UIC program divides injection wells into six classes, based on type. Class II wells "inject brines and other fluids associated with oil and gas production, and hydrocarbons for storage."[9] Given this definition of injection, waste from fracking operations would be pumped into Class II wells. However, the Energy Policy Act of 2005 amended the SDWA definition of "underground injection" to exclude "the underground injection of fluids or propping agents (other than diesel fuels) pursuant to hydraulic fracturing activities related to oil, gas, or geothermal production activities."[10] This infamous exclusion has become known as the Halliburton Loophole because the U.S. vice-president at the time, Dick Cheney, who was instrumental in getting the Energy Policy Act of 2005 passed, had earlier been chairman and CEO of Halliburton. (As we saw in chapter 1, the links between Halliburton and fracking date back to the late 1940s, with the company holding a patent for the technique, which in those days was a wholly vertical drilling procedure.)

Proponents of the Energy Policy Act of 2005 argue that fracking was never regulated under the SDWA in the first place. But if this is indeed true, under what U.S. federal laws is fracking regulated? Moreover, the SDWA sets guidelines or acceptable concentrations for a number of compounds, including many that are components of fracking fluid. But because of the Energy Policy Act of 2005, these compounds became exempt from federal law! This blatant exclusion makes it challenging for citizens to trust that governments are doing their due diligence in protecting water, a distrust further heightened by the secrecy surrounding fracking fluid ingredients.

In the wake of this controversy, a Fracturing Responsibility and Awareness of Chemicals (FRAC) Act was introduced in Congress in 2009, where it still remains, unpassed. The bill would give the USEPA authority over fracking by amending the definition of the UIC program to include the practice. It would also require disclosure of chemicals used. Understandably, industry is generally opposed to the FRAC Act, citing the increased cost per well of adhering to the proposed guidelines. So, while the FRAC Act is stuck in Congress, fracking plows forward with no updated or newly enacted law. One must question the intention behind introducing such legislation and then shelving it for years while water resources are being drained and the environmental destruction and public health risks increase.

At the state level, regulation of fracking is carried out by a number of agencies. Oil and gas regulators oversee many infrastructure aspects of fracking that are put in place to protect groundwater based on local geography and land use characteristics. Many state groundwater agencies are members of the Ground Water Protection Council (GWPC), whose stated purpose is "to promote and ensure the use of best management practices and fair but effective laws regarding comprehensive ground water protection."[11] In conjunction with the Interstate Oil and Gas Compact Commission, the GWPC set up fracfocus.org, the FracFocus Chemical Disclosure Registry, where chemicals used in fracking operations can be voluntarily disclosed well by well. Naturally, the fact that disclosure is voluntary does very little to allay anxiety among concerned citizens.

Currently, North American fracking practices are not governed by any stringent, enforceable statutes to safeguard water from potential threats. Instead, guiding principles and best-practices documents have been put forward by both the Canadian Association of Petroleum Producers (CAPP) and the American Petroleum Institute (API). Keep in mind that the practices set out in such documents are not mandatory, nor are all companies that are involved in fracking even members of these national associations. Two of CAPP's five "Guiding Principles for Hydraulic Fracturing"[12] directly pertain to water:

- We will safeguard the quality and quantity of regional surface and groundwater resources, through sound wellbore construction practices, sourcing freshwater alternatives where appropriate, and recycling water for reuse as much as practical.

- We will measure and disclose our water use with the goal of continuing to reduce our effect on the environment.

Similarly in the U.S., the API publishes a number of such industry guidance documents, as listed in its "Overview of Industry Guidance / Best Practices on Hydraulic Fracturing."[13] The document HF2, for example, "identifies best practices used to minimize environmental and societal impacts associated with the acquisition, use, management, treatment and disposal of water and other fluids associated with the process of hydraulic fracturing." These guiding principles and best-practices documents are rife with motherhood statements pertaining to water management and fracking. In effect they amount to only "soft law," and there is no assurance that any such principles are actually adhered to or taken into account in fracking operations.

Because there are no overarching, enforceable Canadian or U.S. federal laws specifically overseeing fracking or the associated impacts on water, much responsibility for the regulation of these matters is devolved to provincial and state governments. To provide insight into the regulatory challenges of managing water in conjunction with fracking activities, we will look at Alberta and Texas as case studies.

Case study: Alberta

In order to understand water management and how it relates to current fracking regulation in Alberta, we must return to the late nineteenth century, prior to the province's joining Confederation in 1905. The legal framework pertinent to natural resources, which by definition included water, owes its origins to these pre-Confederation circumstances. At the time, under the Constitution Act 1867, the federal government possessed sole jurisdiction over natural resources.

Against the backdrop of the sparsely populated, largely immigrant-driven agrarian economy of the late nineteenth and early twentieth centuries, water was federally governed on the basis of the English common-law riparian rights doctrine. "Riparian" refers to the interface between land and a watercourse, and as the name of the doctrine implies, a right to water use was granted to landowners whose property adjoined a river or stream. However, as settlement in the western territories was encouraged, the population increased and not all landowners were able to settle beside or near surface watercourses – especially in the semi-arid Palliser Triangle, which includes parts of southern Alberta. The need was recognized for a more comprehensive water use law and for the ability to transport water across greater distances.

The federal government responded to these changing water use needs by enacting the North West Irrigation Act (1894). This legislation eradicated the common-law riparian rights doctrine by explicitly stating that all surface water belonged to the Crown. Water rights were allocated by federally issued licences beginning in the late 1800s, contingent on the applicant's demonstration of an appropriate use of water, e.g., irrigation. The licence included such details as volume, maximum diversion rate, source, date etc. An important inclusion to these water rights was that of expropriated land. Water licences were issued under the guarantee of prior allocation, a system that was modelled after western U.S. water allocation regimes. It is important to understand prior allocation, as it continues to govern water licensing in Alberta today.

Prior allocation, or "first in time, first in right" (FITFIR), is a temporal principle; it honours water licences based on their seniority (i.e., date of issuance), with senior (older) licences having priority over junior (newer) ones. Once the licence is issued, its date is the most important determinant in obtaining the quantity of water assigned to the licence, not the intended use or the volume. With this in mind, it is important to note that the water licence "only guarantees the right to take water if sufficient water is available,"[14] a condition dating from long before the current scale of oil and gas development, and changing hydrology, had taken hold in Alberta. From these origins of surface water allocation by means of licensing, water is prohibited from being diverted or used (on anything larger than a domestic scale) without a provincial-government-issued licence. Licences were issued in perpetuity and transferred to successors with the sale of land.

In 1930 the Canadian federal and western provincial governments entered into the Natural Resources Transfer Agreements (NRTA), whereby the governance of natural resources was transferred to the respective provinces by way of a constitutional amendment by the British Parliament (British North America Act, 1930, 20-21 Geo. V, c. 26 [UK]). Under the NRTA, water was only presumed to be included in the definition of natural resources. It was not until 1938, when the Natural Resources Transfer (Amendment) Act modified the original agreement to specifically include water, that the governance of water was unequivocally transferred to the Prairie provinces. Interestingly, despite the uncertainty as to water under the original NRTA, Alberta had already proclaimed the Water Resources Act (1931) to oversee water allocation. Recall that Alberta was predominantly an agricultural economy until the 1950s, with only sporadic oil and gas discoveries. Although the constitutional transfer of water governance had occurred, the federal government did not entirely cede responsibility for water to the Prairie provinces; the establishment of the Prairie Farm Rehabilitation Administration (PFRA) in 1935 ensured the federal government was not without a voice in the water stakes of the Prairies.

Soon after the Natural Resources Transfer

(Amendment) Act, the Alberta government passed the Oil and Gas Resources Conservation Act, also in 1938, to ensure responsible oil and gas development. To see to it that the activities of oil and gas development aligned with the new Act's directives, the independent, quasi-judicial Petroleum and Natural Gas Conservation Board (PNGCB) was established. The PNGCB existed in the same capacity as Alberta's current energy regulatory body, the Energy Resources Conservation Board (ERCB) (2008). Given that these regulatory structures for governing oil and gas development and managing water were put in place long before the onset of Alberta's first significant oil boom, how could they be stringent enough to ensure that petroleum development would occur in a safe manner for generations to come?

In recognition of the need to create updated water legislation, the provincial Water Act was proclaimed in 1999 "to support and promote the conservation and management of water, including the wise allocation and use of water." Interestingly, the Water Act maintained the prior allocation system of FITFIR and allowed for water transfer licences but prohibited both interbasin water transfer and the export of water to the U.S. Without a doubt, the export of water to the U.S. will emerge as a pressing issue in the near future. There are provisions for trade in "resources" under the North American Free Trade Agreement that will undoubtedly be put into effect if the current water use and management practices surge forward in their present state of overuse and overallocation. This leads to the consideration of water as a commodity. Water scarcity in southern Alberta (not to mention the southwestern states) is recognized as a pressing threat to ecosystem integrity, agriculture and energy development. It is important to note that unconventional resource extraction is being carried out in many of these increasingly semi-arid to arid regions. Alberta considers itself a water-endowed province, which relatively speaking it is, but that is quickly changing, due in part to unconventional resource extraction. However, water management in the province remains static in its outdated practices.

An important mandate that emerged from the Water Act was the development of a provincial framework for water management. The Water for Life Strategy (2003) was instituted with three overarching goals supported by assurances to Albertans that: they will have a safe, secure drinking water supply; healthy aquatic ecosystems will be maintained; and reliable, quality water supplies will be managed to support sustainable economic development. On paper these goals and assurances are meant to instill confidence that provincial water management is thoughtfully considered and proactive. However, in light of accelerating resource development and emerging global environmental threats, it is difficult to pinpoint how exactly these goals are being achieved and enforced.

First, although safe drinking water is ensured for the majority of the provincial population, a large cohort is outside the jurisdiction of drinking water regulations because they draw water from private or domestic wells. No overarching federal law oversees drinking water quality; only guidelines are in place. Second, the federal government has power over the integrity of aquatic ecosystems by way of the Fisheries Act, so Alberta must comply with federal provisions regardless of its own stance on aquatic ecosystem integrity. Lastly, it has been acknowledged that waters from certain large basins in the province's south have been overallocated since 2006, when the transition to unconventional resource extraction was in its infancy. Water is vital to grain and livestock production in those locales. How are we going to contend with already overallocated basins, continued water-intensive resource abstraction and growing human and livestock populations, knowing that our water sources have long been overburdened? Even though these basins are known to be overallocated, there has been no dramatic, visible decrease in the amount of water flowing across the surface. Perhaps this lack of any visual cue serves to justify the inaction of government and regulatory bodies to amend water management plans in light of increasing pressures.

In the vein of stewardship, the Water for Life Strategy was the impetus for establishing provincial Watershed Planning and Advisory Councils (WPACs). Presently there are eleven WPACs at various stages of developing and publishing comprehensive "State of the Watershed" reports as a precursor to "Integrated Watershed Management Plans." The WPACs are multi-stakeholder, independent, non-profit organizations created in part to address local watershed concerns through consensus-based decision-making while fostering collaborative interactions among the various stakeholders, which may include industry, government, Aboriginal communities and watershed residents. The most recent WPAC is the Mighty Peace Watershed Alliance (MPWA), established in 2011. The MPWA is concerned with Alberta's largest watershed, which drains one-third of the province northeastward toward the Mackenzie, Canada's longest river.[15] Despite being the newest WPAC, the MPWA must contend with the threat of intensive fracking operations. Further assaults to the watershed include large-scale land use changes resulting from other intensive resource development.

Under Alberta's Water Act, original water licences are still in effect under the FITFIR system of prior allocation, which, as previously mentioned, comes into effect in periods of water shortage. As the stressors on Alberta water sources mount, a contingency plan is needed to ensure there is enough water for the other Prairie provinces (pursuant to the 1969 Master Agreement on Apportionment), for irrigation in Alberta's southern breadbasket and for domestic and municipal use. It is important to remember that not only fracking taxes Alberta's water resources; so too do the oil sands, climatic changes and population growth.

The ERCB is the regulatory body charged with issuing permits and approvals for oil and gas operations today. Numerous directives are in place that directly relate to fracking, many of them focusing on the engineering and infrastructure of wells. However, there is no specific regulation on the distribution density of wells, that is, the number of wells per unit of territory. This is problematic in light of the extent and intensity with which wells are being fracked. Moreover, a potential water contamination pathway arises from intersection with previously drilled or abandoned wells, hundreds of thousands of which honeycomb the subsurface terrain. Logically, as the density of wells increases so too does the likelihood of intersection with other wells, new or old. Furthermore, it is difficult to ascertain whether permits are considered and issued based on status quo "normal operations" or worst-case scenarios. This is a vital consideration, knowing that when it comes to groundwater protection, the bulk of the regulatory scrutiny is on the engineering and design of well bores.

Obviously, when the system of prior allocation was implemented in the late 1800s, water scarcity, climate change and the scale of current oil and gas development could not have been estimated or planned for. Yet these outdated water management approaches remain in effect. The management of water based on presumed abundance must adapt to contend effectively with the pressures on our dwindling water resources. From changing hydrologic regimes to the massive water volumes required for extracting unconventional oil and gas, Alberta's water sources are under assault. The province is at a critical point and must put water at the forefront of its concerns, policies and frameworks. It is not good enough to simply create policy as events arise that compromise water sources; the province must be proactive in developing and implementing its policies and must follow through with stringent actions to safeguard our water before irreversible damage occurs.

Case study: Texas

Texas water law, like its Alberta counterpart, owes its origin to English common law. There are, however, striking contrasts in water governance between Alberta and Texas, especially as it pertains to groundwater allocation. Texas water law is split among various levels of government and regulatory agencies, with exclusions for oil and gas activities that mirror policies and regulations implemented at the federal level. In Texas, surface water is owned and managed by the state, whereas groundwater belongs to the landowner.

Surface water accounts for less than half the overall water use in the state. Withdrawals or diversions by individuals, corporations or cities require a state-issued permit granted by the Texas Commission on Environmental Quality; water right permits are valid in perpetuity. Under the Texas Water Code, exemptions to the requirement for state-issued permits include domestic and/or livestock use, wildlife management, emergencies and other specified uses.[16] As in Alberta, surface water allocation in Texas was founded on the English common-law riparian rights doctrine. Surface water in Texas was also governed by the doctrine of prior allocation, which mirrors the system regarding prior allocation described in the Alberta case study above. In 1967 the doctrines were merged as a result of the enactment of the Water Rights Adjudication Act, which in effect unified the water permit system as it pertains to surface water allocation.[17] Akin to the situation in semi-arid southern Alberta, many surface water bodies in Texas became overallocated decades ago. Many permits have not drawn their full potential of water up until now, but this is bound to change in the wake of the droughts of 2011–12, population growth and statewide industrial pressures on freshwater resources. Similarities of allocation governance between Alberta and Texas end with surface water rights, however.

In Texas, as we've said, groundwater is the primary source; much of the state's economy, prosperity and health hinge on it. And because groundwater is owned by the landowner, it is governed as a private property right using the rule of capture, which stems from the English common law of "absolute ownership." It is important to note that absolute ownership was also applied to the ownership of oil and gas by the Texas Supreme

Court. The rule of capture was first explicitly adopted by the state Supreme Court in 1904 and is still adhered to today despite pleas for a more equitable and just system of water allocation, especially from private/domestic well owners. Under the rule of capture, landowners own the groundwater underlying their property and possess unlimited access to the resource, even if it adversely affects a neighbouring well. The rule of capture imposes no limits on the amount of water that can be withdrawn from beneath the land for personal use or sale. Neither does it enforce any liability if a landowner depletes or exhausts a neighbouring well, nor does it require the extracted groundwater to be used on the overlying land. And, of the utmost importance in semi-arid to arid Texas, the rule of capture does not require the landowner to reduce pumping during water shortages! We know droughts are becoming increasingly common in that state (not to mention the entire U.S.) as a result of climatic changes, so how can there be no enforceable limits on the volume of water that can be abstracted during times of dryness or drought? The summer of 2011 was the hottest and driest in the meteorological history of Texas, with the vast majority of the state classified in the two worst categories of drought: "extreme" and "exceptional."[18] It does not seem a stretch to imagine that Texans will have to anticipate and be prepared for more drought years ahead.

Given the strictures of the rule of capture, it is obvious why all major states (save Texas) that had adopted it have evolved their groundwater management practices toward systems that are more equitable and just. In more than a century of practising the rule of capture, Texas landowners relying on groundwater have made several attempts in court to have the groundwater allocation system updated, but to no avail. Undoubtedly, as fracking operations proceed with gusto in the Lone Star State, there will be an increase in lawsuits and claims of injustice owing in part to the antiquated rule-of-capture system, which leaves landowners at the mercy of their neighbours and is colloquially referred to as the "law of the biggest pump." When your neighbour becomes a fracking company, there is no legal recourse to ensure that your water quality or quantity is safeguarded. This is frightening, knowing the current and projected pace of unconventional extraction and the quantities of fresh water required for coaxing subterranean unconventional resources out from beneath Texas.

In response to a groundwater well interference case in which the rule of capture was upheld by the Texas Supreme Court, the state authorized the establishment of local Groundwater Conservation Districts (GCDs) empowered to "promulgate rules for conserving, protecting, recharging, and preventing waste of underground water."[19] Notably, withdrawing groundwater for petroleum activities is exempt from the Texas Water Code – a clear demonstration of the state's pro-oil-and-gas position, which explicitly supersedes groundwater preservation and protection. There are no attempts by regulatory bodies to control water use, and there are no consequences for industry in using as much water per well as it deems necessary. This needs to change before freshwater resources become taxed beyond recovery.

Because groundwater is so important to the sustainability of Texas, numerous authorities (in addition to GCDS) are involved in groundwater management: river and aquifer authorities, water utilities, municipalities, counties and the

Texas Water Development Board. With so many influences on various aspects of groundwater, it is challenging to pinpoint specific roles and responsibilities. Liability for any adverse effect on groundwater cannot be clearly applied to any one authority or body. This is inherently problematic when groundwater is compromised in any way. It is important to note that of all the stakeholders in groundwater governance, no single body has any legal authority or recourse over groundwater infractions that may result from oil and gas activities. Given the extent and intensity with which wells of all types are drilled in Texas, it is astonishing that there is no system of checks and balances when it comes to water use. What protective purpose do the groundwater bodies fulfill if they possess no legal authority over the resource?

The regulatory agency charged with overseeing Texas's prolific oil and gas industry is the Railroad Commission of Texas (RRC), established in 1891. As the name implies, its inception was linked to the railway industry, but the commission quickly expanded its mandate to include oil and gas. Today the RRC is involved with all aspects of petroleum development. With regard to fracking operations, companies are legally required to report the amount of water used; however, information on the source and quality of the water is not required,[20] making the assessment of water resources nearly impossible. As we know, water can be obtained from surface or groundwater sources or a combination thereof, in addition to full or partial use of saline or brackish water. It has been noted that the RRC allows petroleum companies to use as much water as necessary to complete a well. No water conservation measures are in place, nor any deterrents from the regulators that should hold companies accountable for their water-intensive operations. As in other arid and semi-arid regions of the world, freshwater resources in Texas are dwindling. What is unique about Texas is its burgeoning unconventional sector (fracking having been first carried out in the Barnett play), amplified by the maintenance of the rule of capture and governance at all levels that favours the oil and gas industry.

Because Texas is the bellwether of fracking and maintains prominence in all aspects of the oil and gas industry, scientifically robust studies of environmental changes that arise from fracking, as well as literature on all aspects of the practice, have been most plentiful in the Texan context. States, provinces and countries have their eyes on Texas to observe how development proceeds and how regulators tackle fracking amid increasingly loud public concerns. This is not the example we want to follow. Governance in Texas clearly favours oil and gas and has long forgone environmental laws and regulations. Although Texas has paved the way for unconventional resource extraction, allowing us all to continue living with the conveniences we rely on, this is an opportunity to set ourselves apart by scaling back the pace of unconventional development to allow for the invention of less water-intensive extraction technologies and perhaps the further exploration of fuel sources that do not depend on hydrocarbons.

At first glance, it appears as though extensive consideration has been given to watershed management and water conservation measures right alongside robust petroleum economies in Alberta and Texas. However, upon analysis of the practical, on-the-ground activities, it quickly becomes apparent that although water is a formal consideration for regulators, oil and gas take precedence. Water is not protected from intensive fracking operations. It is being drained right out from under us, and what water is left behind is becoming increasingly contaminated. Because the U.S. and Canada are leaders in the unconventional resource development revolution, the onus is on their federal governments to set a global precedent for safe, sustainable extraction while ensuring long-term availability of fresh water for all citizens. To date, neither government has done so. Unconventional resource extraction surges forward at unprecedented rates despite known adverse environmental and public health effects.

Environmental and Health Concerns

With water issues at the forefront, concerned citizens and landowners have sounded the alarm over the numerous public health and environmental problems that arise from fracking. In some instances, such objections have been heeded, with moratoria or even all-out bans implemented in order to gather sufficient information to make an informed decision. This is far from the norm, however, and frustration and dissension are mounting among concerned citizens as fracking proceeds at unprecedented rates without due consideration for any misgivings raised. Very real public health and environmental threats are associated with unconventional fossil fuel extraction, and these need to be properly addressed before irreparable damage is done. We have all been subjected to the media's fear-mongering tactics on a wide variety of issues. What makes fracking any different? It is simple:

we are talking about the potential devastation of potable water sources with limited to non-existent recovery potential. We cannot survive without water! All the water that will ever be on Earth is here today. What has changed is that water is being permanently removed from the hydrologic cycle at accelerated rates, in large part due to intensive unconventional resource extraction.

The term "ecotoxicology" was coined in 1969 by the discipline's founding father, René Truhaut, to identify the branch of toxicology concerned with studying the toxic effects of pollutants to the constituents of ecosystems. It is a tenet of toxicology that "the dose makes the poison." In other words, the amount of substance ingested determines its toxicity and resultant health effects. There are deep-seated concerns pertaining to the ecotoxicological effects of fracking fluid components. When these compounds make their way into freshwater resources, they may bioaccumulate, biomagnify, persist and permanently alter aquatic ecosystem integrity, ultimately adversely affecting human health. Of the hundreds of identified fracking fluid components that have been voluntarily disclosed by companies or elucidated by scientific studies, many produce well-documented adverse health and ecosystem effects.

One example is benzene. Benzene was documented in *Silent Spring* as a compound which "lodges in the [bone] marrow and remains deposited there for periods known to be as long as 20 months. Benzene itself has been recognized in medical literature for many years as a cause of leukemia."[1] There is consensus among renowned medical and environmental organizations that benzene is a human carcinogen. Yet the chemical is used in some fracking operations. Furthermore, according to the USEPA, benzene is regulated under the SDWA with a maximum contaminant level goal (MCLG) of zero (non-enforceable), and a maximum contaminant level (MCL) of five parts per billion (enforceable).[2] However, we must return to the Energy Policy Act of 2005 to remind ourselves that fracking operations are exempt from the SDWA. Therefore, despite the MCLG and MCL put in place by the USEPA to safeguard citizens from the known adverse effects of certain compounds, including benzene, there is no protection from indiscriminate chemical use in fracking operations. This is appalling in light of the well-known, well-documented, well-accepted carcinogenic nature of benzene, only one compound out of the hundreds that have been identified! How can citizens be expected to accept that the current regulatory structure is adequate to ensure public health and environmental welfare in the face of such blatant legislated immunity from a powerful, overarching federal law? And given that, to date, disclosure of fracking fluid components is only carried out on a voluntary basis, how can adequate management plans be designed to incorporate known ecotoxicological effects of fracking fluid components when only partial transparency (at best) is in place?

The acute and chronic consequences of many individual fracking fluid components have been documented. What remain elusive are synergistic effects that may arise as a result of injecting the fracking fluid cocktail, i.e., interaction among components of the fluid and interaction of fluid components with extraction infrastructure and local geology. It is relatively simple to isolate compounds, chemicals and ingredients to demonstrate their immediate or long-term health or environmental effects. But it requires a longitudinal study to gather the necessary information to inventory the evidence and gain an understanding of synergistic effects. Given the limited study of current fracking practices, it is not possible to have gathered sufficient data to understand long-term, chronic and/or synergistic effects of many fracking fluid components. This realization is frightening in light of the current pace of unconventional development. We need more time to gather the information, process it, understand it, integrate it and then move forward in a sustainable, rational manner conducive to long-term preservation of our ecosystems and well-being. We must have a better grasp on what we are inflicting on ourselves and the environment before we cause irreversible destruction.

In order to enhance the recovery of oil from conventional reserves, studies of microbial enhanced oil recovery (MEOR) have been carried out for years.[3] MEOR has met with limited success, owing in part to its largely laboratory-based investigation, with little opportunity or demand for real-world testing. Partially due to the seamless transition from conventional to unconventional extraction, we have not reached the critical point of absolute petroleum shortage at which enhancement technologies such as MEOR would be vital. In order to bioengineer micro-organisms, the naturally occurring consortia – planktonic and biofilm – must be well understood to ensure efficacy of enhancement. However, only one or two studies to date have attempted to characterize microbial communities in deep subsurface plays that are being fracked.[4] Engineering of micro-organisms that enhance shale gas extraction is further complicated by the addition of biocides to the fracking fluid cocktail in order to control microbial growth that is naturally present in water and geologic formations. Numerous biocides are used to limit corrosion of well infrastructure caused by naturally occurring microbial processes, but these have proven inadequate to kill all micro-organisms present in fracked wells. What's more, biocides, often chemical in nature, impose additional sources of ecotoxicological stress that may compromise ecological integrity and public health. Given the marriage of petroleum and bioengineering disciplines at numerous North American institutions, it is safe to assume that financial resources are being allocated toward gaining a better understanding of how to manipulate natural microbial processes to enhance shale gas extraction. It is interesting to note that this unsustainable branch of science is being fostered while research into alternatives to fresh water use in fracking is not garnering the same respect or financial endowment.

What must also be addressed are the fracking fluid components that remain underground. Recall that the optimistic estimate is that 80 per cent of flowback and produced water returns to the surface over time. This means that at least 20 per cent of the injected concoction remains underground, where harmful components may migrate to the surface or intersect with groundwater aquifers via the hydrologic network over time. Not knowing what chemicals are there, or when or where they might re-enter our potable water system, makes it exceptionally challenging to anticipate their effects and plan accordingly. Contending with the chemicals that do return to the surface poses another set of challenges. Can our water and wastewater treatment facilities, as they presently operate, remove chemicals or ensure their concentrations are at safe levels before the treated water is returned to the hydrologic system? This is doubtful in light of reports by national organizations such as the Canadian Public Works Association and the Federation of Canadian Municipalities on the poor state of our nation's wastewater infrastructure.[5] As mentioned earlier, fracking is outside the ambit of federal water laws in Canada and the U.S., and there is no complete inventory of what is being injected and subsequently returned to the surface. This lack of transparency makes it impossible to mitigate any risk or disaster effectively enough to ensure the integrity of our water resources. And then there are all the private water sources that are not monitored, regulated or accounted for under any drinking water laws. These are only a few of the areas of concern that arise when considering the assaults fracking is imposing on our water systems and health. We must also contend with another, even greater unknown: climate change and its impacts on hydrology.

Just how climate change affects hydrology depends on the geographical location of the water body. What we are generally experiencing in North America in the high and mid-latitudes are increased precipitation events (flooding), whereas the southern latitudes are experiencing increasingly arid (drought) conditions. Although climate change models are still rudimentary and the projected changes to hydrology cannot be fully predicted, we know that our already taxed water systems are further pressured by these increasingly intense changes to the global climate. Demands for water are increasing while its quantity is diminishing. This has already led to bidding wars in the last few years between the agricultural and oil and gas industries in Texas and Colorado.[6] Both are vying for water allocations in increasingly arid conditions. It can only be assumed that this scenario will play out more often and in more places as our hydrology changes and unconventional extraction continues apace. The issue of water quantity and fracking is gaining momentum in mainstream media as the U.S. faces the most severe and widespread drought recorded in the past fifty years.[7] Undoubtedly, this phenomenon will repeat itself in Canada as the number of fracking operations increases.

The knowledge that hydrology is changing should be enough to prompt the modernization of outdated water management systems in favour of protecting water by whatever means possible. While it is understood that there are inextricable ties between energy and water, we must implement a system whereby both are preserved, protected and sustained for years to come. We are facing uncertainty – of climate, of hydrology, of the long-term health and environmental effects of actions being carried out today – and yet there seems to be no deterring this unconventional resources revolution. We need to invest time, intellect and money to find alternatives to massive freshwater use in fracking operations, since potential substitutes such as saline or brackish water are currently cost-prohibitive to treat. This is the only sustainable solution. Undoubtedly such measures will require scaling back unconventional extraction in the interim, which in turn will reduce the amount of water consumed by industry. There is also a need to put more stringent water and energy conservation measures in place at both industrial and domestic scales. Our water and energy are too cheap; there is little to no motivation to conserve when energy and water prices are hardly noticeable.

Groundwater

"Groundwater" is a generic term applied to water that is ubiquitous beneath the Earth's surface. Although it comprises less than 1 per cent of the world's available fresh water, groundwater supplies drinking water to millions of people around the world in addition to supplementing food production. It is harboured predominantly in impervious soil, sediment and/or rock pores, but can also collect in pools known as aquifers. These properties of groundwater are akin to those of fossil fuels, but the obvious contrast between the two is that we cannot live without water. Somehow this fundamental truth has long been overlooked in water management frameworks; groundwater's subsurface situation has translated into its frequent exclusion from water management frameworks – out of sight, out of mind!

According to a common misconception,

groundwater exists as massive subterranean watercourses analogous to rivers on the surface. However, this is not the case. If groundwater moves, it moves slowly, not in rushing torrents that carve out new pathways beneath the Earth's surface. These events are slow and occur over thousands of years. Perhaps this false image of subsurface rivers contributed to groundwater's exclusion from water management frameworks for so many years – yet another way in which our water resources have been administered on the erroneous basis of presumed abundance. But now we know better, and the time is upon us to update our stewardship to give groundwater the same importance as readily visible surface water has always had.

Groundwater is becoming increasingly important as a secure potable water source as surface waters dwindle and hydrology changes. In fact, many regions of the southern U.S. rely solely on groundwater for their potable supplies. Coincidentally, in many of these same regions, intensive fracking operations are tapping into plays at various depths and proximities to groundwater sources. This should be of great concern to citizens and decision-makers alike. Amplifying these concerns is the realization that with current technology it is nearly impossible to remediate contaminated groundwater. The techniques that have been identified are generally cost-prohibitive, if they are even applicable in real-world settings. At a bare minimum, if we are going to continue exploiting groundwater in favour of fossil fuel development, we need to carry out intensive research to develop sound groundwater remediation techniques in anticipation of contamination events. Because contamination will occur; it is inevitable.

As noted earlier, water is interconnected, and thus groundwater is recharged by precipitation and runoff. What is important to understand with source groundwater is that not only do large aquifers supply entire regions, districts or counties with potable water, but millions of private wells also derive their water, for domestic and small-scale agricultural use, from aquifers. As also previously mentioned, these private wells are beyond the regulatory scope in Canada, and in the U.S. they are exempt from the provisions of the SDWA. This is a huge regulatory gap, as the bulk of fracking operations occur in rural settings.

Furthermore, recall that the rule of capture employed in Texas legally entitles a landowner to exhaust a neighbouring well as long as no malice was intended. This does not bode well for preserving groundwater supply in rural settings.

To date, our management of groundwater has been severely lacking. Baseline data and monitoring programs to track changes – whether anthropogenic or natural – to groundwater courses and aquifers are limited. We are missing a sufficient inventory of all aspects of groundwater: quantity, quality, situation, delineation of aquifers. What's more, we lack insight into the direction in which groundwater flows and the effects of these flow regimes on surface waters; recall the interconnectivity of water resources. This is a concern in light of current unconventional oil and gas development in increasingly close proximity to, or directly through, groundwater aquifers. The potential threats to groundwater from fracking include contamination from leaching fluids, inadvertent intersection with previously drilled wells, depressurization of aquifers, and acute or chronic changes to water quality resulting from aquifer contamination. Moreover, claims of groundwater contamination have cited such causes as old, inadequately plugged abandoned wells; poorly constructed, cemented or sealed wells; overpressurized wells; and wastewater pit leaks.[1] Other well-known stresses on groundwater resources include the aquifer-mining and overdrawing that arise from population growth and changes to local hydrology.

When the vertical section of a well is drilled, it pierces groundwater aquifers. Although the destination of the well bore or the spot where it angles from vertical to horizontal usually lies below clean groundwater aquifers, the bore nevertheless passes through them. Colloquially, there are two types of groundwater: good and bad. Good groundwater is generally used for potable water and found at shallower depths, whereas bad groundwater is found at greater depths and is often saline and/or contains large amounts of total dissolved solids. Owing to the chemical nature of bad groundwater, it is not considered economically viable to clean it for use as a potable water source.[2] Shale gas fracking operations are located in and around bad groundwater aquifers; however, they must pass through the good groundwater aquifers, and the extraction of

CBM occurs at depths aligned with those of good groundwater aquifers.

Given that the well bore pierces several strata comprised of various rock, water and other geologic formations, cementing the well is complicated – honeycombing, or holes in the cement, can occur. The consequent gaps serve as a potential migration path for contaminants into the surrounding geology, from which they may then find their way into groundwater aquifers. These well bores also cut through natural filtration layers comprised of diatomaceous earth (DE). Layers of DE form over millions of years and are comprised of the algae, called diatoms, that have siliceous cell walls. DE layers function as a subterranean natural filtration medium, analogous to surface wetlands, which aids in the natural purification of groundwater. Disrupting these layers has many adverse effects, one of which is a reduction in groundwater cleansing.

As our surface waters, which were once perceived as abundant, reach a tipping point from which they cannot recover in a timely manner, we will be moving more rapidly and to greater depths as we pursue groundwater resources to substitute for the diminishing surface water. We do not have a complete understanding of our groundwater resources; we have excluded them from the water management equation for too long. We need to amend this now, through the development of robust groundwater management frameworks. As James P. Bruce iterates in his eloquent report "Protecting Groundwater: The Invisible but Vital Resource," the basis for sound, sustainable groundwater management frameworks is protection from depletion and contamination; protection of ecosystem viability; allocation to maximize groundwater's contribution to economic and social well-being; and application of good governance.[3] These objectives can be met through the development of groundwater resource inventories, baseline studies and comprehensive monitoring programs. They can be achieved; they just need to be addressed and pushed to the forefront of the regulatory agendas.

The issue of groundwater management is time-sensitive. We must act now and we must be vigilant in our quest to develop and implement these proactive frameworks. We are going to need groundwater resources sooner and in greater abundance than we have projected. Let's not allow fracking operations to compromise vital groundwater resources when we know better.

Future Focus

If there is today a healthy degree of public awareness about water security and water wars, this was not always the case. In 1995, when a World Bank vice-president for environmentally sustainable development, Ismail Serageldin, publicly stated his prediction that the wars of the twenty-first century would be about water,[1] the notion was scoffed at and dismissed as being alarmist. Fast-forward to today, however, and the possibility that Serageldin may indeed have been correct has permeated several facets of North American culture. In the past few years, an increasing number of people have asked me if I am aware that the next world war will be fought over water. This is an indication of an enhanced level of knowledge and an acceptance of the importance of water in a global context. This acknowledgement of the crucial role water plays in all of our lives is vital to its preservation. We do not protect what is not important to us. For too long we have ignored water, taken it for granted. It has been managed on presumed abundance, to the detriment of the quality and quantity of our freshwater resources. This must change.

As was demonstrated in the 1960s and '70s, environmental change was driven in part by grassroots movements across nations. We North Americans, as citizens of developed democratic nations, have the power to effect the wave of change that is necessary to protect our freshwater resources from the assaults of fracking. We must all actively champion the protection of our water; we all have a stake in preserving its integrity. How do we participate? First, we must educate ourselves and use knowledge as power. We must ensure that fracking and fresh water are included in election campaigns and debates. Electing officials with water conservation objectives will at a minimum put water on governmental agendas – a vast improvement from recent election platforms. Our politicians must be held accountable to their promises of ensuring safe, clean water for generations to come. And we must educate our children. Children are one of the most effective resources we have for creating change in our society. We must engage our youth and children by fostering a sense of connection to our natural environment. When we are connected, we are more apt to take a vested interest in causes and issues that threaten our well-being.

Initiatives across the continent aim to scale back the development of unconventional fossil fuels and ensure that our environment and public health, not resource extraction, are of primary importance. Because the U.S. is the cradle of fracking, the world is looking to it for guidance and precedents. Several states are drafting and implementing stricter legislation around shale gas development.

In June 2012 the governor of Ohio signed precedent-setting oil and gas regulatory legislation. Included in the statute is the nation's first public chemical-disclosure registry that covers companies throughout their entire lifespan, from start-up to dissolution. The existing federal and state "trade secret" laws remain in effect, but the state can obtain proprietary information in the event of an investigation or spill. Not complete transparency, but getting closer. Disclosure of chemical information to physicians is mandatory, and doctors may share even proprietary information with patients and other medical professionals where necessary. Additionally, the source and volume of water and the rate at which it is to be used throughout the fracking operation must be disclosed.[2] Recall that in Texas, disclosure of the source is not required. Ohio is a prime example of a state that has experienced moderate oil and gas development (albeit long-term) and is now on the brink of full-force unconventional extraction. Both the Utica and the Marcellus plays underlie Ohio.

Vermont has banned fracking. Vermont is interesting because it is not typically a gas-producing state, although like all of North America it reaps the benefits of fossil fuel production.[3] The state's ban is criticized as being merely symbolic because of the lack of substantial hydrocarbon reserves; however, the ban demonstrates proactive leadership ahead of potential unconventional resource development. Similarly, North Carolina, which does have proven unconventional reserves, is demonstrating the precautionary principle and proceeding with proactive legislation to restrain the pace of shale gas development in the state. As governor Beverly Perdue, a Democrat, stated, "Our drinking water and the health and safety of North Carolina's families are too important. We can't put them in jeopardy by rushing to allow fracking without proper safeguards."[4] Other states and provinces would be well advised to follow this kind of leadership. The unconventional resource revolution is making inroads into "new," non-traditional oil and gas producing areas, and because many such states and provinces lack a regulatory structure aimed at managing oil and gas development, the opportunity is there to develop proactive legislation.

In Canada, British Columbia is setting precedent for regulating the fracking industry. FracFocus.ca, the Canadian equivalent of the U.S.'s online chemical disclosure registry, was launched in January 2012, and under BC legislation the "public disclosure of ingredients used for hydraulic fracturing" is now mandatory. Following suit, Alberta has announced its implementation of enforceable public disclosure of fracking ingredients for 2013. There is a moratorium on fracking in Quebec, as previously mentioned, and a full ban on fracking in Nova Scotia until 2014.

There is a staggering amount of literature pertaining to water management in North America. A leading example of policy associated with a robust natural resource economy that recognizes the intrinsic value of water is the Government of the Northwest Territories' *Northern Voices, Northern Waters: NWT Water Stewardship Strategy*. Developed under the premise of collaborative partnerships in Canada's North, the strategy puts forward five guiding principles for watershed stewardship: respect, sustainability, responsibility, knowledge and accountability – guidelines that water policy-makers would be well advised to incorporate into proactive, effective water policies at any level of government.[5] At a local level, the city of Dawson Creek, BC, provides a proactive model for balancing local oil and gas resource development, including fracking, with long-term water conservation objectives. Dawson Creek has installed an advanced wastewater recycling facility that produces water to be used in part in local fracking operations.[6]

By encouraging water policy reform and

conservation initiatives across all jurisdictions, we can effect positive, necessary change to safeguard our precious freshwater resources. In order for these mandates to be credible, they must be informed by robust scientific evidence and they must reflect concerns voiced by the public. We must be steadfast and vigilant to achieve our water preservation objectives in the face of extensive fracking operations across the continent. The power resides within each one of us and in all of us together.

Endnotes

Introduction and Chapter 1

1. Robert D. Bott, *Evolution of Canada's Oil and Gas Industry* (Calgary: Canadian Centre for Energy Information, 2004): 15.

2. Carl Coke Rister, *Oil! Titan of the Southwest* (Norman: University of Oklahoma Press, 1949): 1.

3. Mark Jaccard, "Peak Oil and Market Feedbacks: Chicken Little versus Dr. Pangloss," in *Carbon Shift: How the Twin Crises of Oil Depletion and Climate Change Will Define the Future*, ed. Thomas Homer-Dixon (Toronto: Random House Canada, 2009): 97.

4. J. David Hughes, "Hydrocarbons in North America," in *The Post Carbon Reader: Managing the 21st Century's Sustainability Crisis*, eds. Richard Heinberg and Daniel Lerch (Healdsburg, Calif.: Watershed Media, 2010), accessed (pdf) April 5, 2013, www.postcarbon.org/Reader/PCReader-Hughes-Energy.pdf.

5. Wilf Gobert, "The Changing Face of the Canadian Oilpatch," Canadian Society of Geophysicists *Recorder* 29, no. 3 (March 2004): 46–47, accessed (pdf) April 5, 2013, via link from http://is.gd/v4KYdE.

Chapter 2

1. Tim Probert, "Shale Gas Fracking: Water Lessons from the US to Europe," *Water & Wastewater International* 27, no. 2 (April/May 2012): 22–24, accessed April 5, 2013, http://is.gd/4Gw28V.

2. Dianne Rahm, "Regulating Hydraulic Fracturing in Shale Gas Plays: The Case of Texas," *Energy Policy* 39, no. 5 (May 2011): 2974–81.

3. Heather Cooley and Kristina Donnelly, *Hydraulic Fracturing and Water Resources: Separating the Frack from the Fiction* (Oakland, Calif.: Pacific Institute, June 2012): 4, 6, accessed (pdf) April 5, 2013, www.pacinst.org/reports/fracking/full_report.pdf.

4. Joe Carroll, "Worst Drought in More Than a Century Strikes Texas Oil Boom," *Bloomberg News*, June 13, 2011, accessed August 6, 2012, http://is.gd/1E0EBF.

5. P.C.D. Milly et al., "Stationarity Is Dead: Whither Water Management?" *Science* 319 (February 1, 2008): 573–74, accessed (pdf) April 5, 2013, http://is.gd/2QUZq1.

6. Abrahm Lustgarten and ProPublica, "Drill for Natural Gas, Pollute Water," *Scientific American*, November 17, 2008, accessed April 5, 2013, http://is.gd/IdWxbc.

7. Bernard D. Goldstein, Jill Kriesky and Barbara Pavliakova, "Missing from the Table: Role of the Environmental Public Health Community in Governmental Advisory Commissions Related to Marcellus Shale Drilling," *Environmental Health Perspectives* 120, no. 4 (April 2012): 483–86, accessed April 5, 2013, http://is.gd/KO32Zb.

Chapter 3

1. Jonathan A. Patz et al., "Climate Change and Waterborne Disease Risk in the Great Lakes Region of the U.S.," *American Journal of Preventive Medicine* 35, no. 5 (November 2008): 451–58, accessed (pdf) April 5, 2013, http://is.gd/zVMJ2w.

2. Canada Water Act, RSC 1985, c. C-11, long title, http://canlii.ca/t/hz6l;
   Fisheries Act, RSC 1985, c. F-14, ss. 34–42, http://is.gd/VOYra1;
   Canadian Environmental Protection Act, 1999, SC 1999, c. 33, http://canlii.ca/t/51xdb; all three accessed April 5, 2013.

3. Environment Canada, *Federal Water Policy*, 1987, accessed August 19, 2012, http://is.gd/i6bNxS.

4. Environment Canada, *Toxic Substances Management Policy*, 1995, accessed August 19, 2012, http://is.gd/YAzMGb.

5. Prairie Provinces Water Board, Master Agreement on Apportionment, 1969, as amended, current to 2009, Scheds. A (Alta./Sask.), B (Sask./Man.), accessed August 5, 2012, www.ppwb.ca/information/79/index.html.

6. USEPA, "Water: Laws & Executive Orders," 2012, accessed July 29, 2012, http://water.epa.gov/lawsregs/lawsguidance.

7. USEPA, "Safe Drinking Water Act, 1974 (SDWA)," 2012, accessed July 30, 2012, http://is.gd/ri1jXs.

8. Environment Canada, "What Is the Great Lakes Water Quality Agreement?" 2012, http://is.gd/xSZ90U;
   USEPA, "Great Lakes Water Quality Agreement," 2012,

www.epa.gov/glnpo/glwqa. See also John Jackson and Jane Elder, "Frequently Asked Questions: Great Lakes Water Quality Agreement Renegotiation 2011–2012," (Kitchener, Ont., and Buffalo, N.Y.: Great Lakes United, 2011), via link from http://is.gd/7ssNqw. All three accessed April 5, 2013.

9. USEPA, "Classes of Wells," 2012, accessed April 5, 2013, http://water.epa.gov/type/groundwater/uic/wells.cfm.

10. Energy Policy Act of 2005 [U.S.], Pub. L. 109–58, sec. 322, http://is.gd/KUwXXQ. See also USEPA, "Hydraulic Fracturing Background Information," http://is.gd/jwCST4; "Regulation of Hydraulic Fracturing under the Safe Drinking Water Act," http://is.gd/JACod6. All three accessed April 5, 2013.

11. Ground Water Protection Council, "GWPC History," 2012, accessed August 6, 2012, www.gwpc.org/about-us/gwpc-history.

12. Canadian Association of Petroleum Producers, "Guiding Principles for Hydraulic Fracturing" (Calgary: Canadian Association of Petroleum Producers, 2011), accessed (pdf) April 5, 2013, via link from http://is.gd/s9LymC.

13. American Petroleum Institute, "Overview of Industry Guidance / Best Practices on Hydraulic Fracturing" (Washington, DC: American Petroleum Institute, 2012), accessed (pdf) August 19, 2012, http://is.gd/MpLuSg.

14. Canada West Foundation, "The Evolution of Water Policy in Alberta" (Calgary: Canada West Foundation, 2010): 2, accessed (pdf) March 3, 2013, http://is.gd/kmMp9m.

15. "About MPWA," Mighty Peace Watershed Alliance, accessed August 24, 2012, www.mightypeacewatershedalliance.org/about-mpwa.

16. Texas Water Development Board, "A Texan's Guide to Water and Water Rights Marketing" (Austin: Texas Water Development Board, 2003): 6, accessed (pdf) April 5, 2013, http://is.gd/6AxraX.

17. Ibid.

18. Justin Petrutsas, "Comparing the 2011 and 2012 Texas Droughts," College Station Weather Forecast, *Houston Weather Examiner*, July 26, 2012, accessed April 5, 2013, www.examiner.com/article/comparing-the-2011-and-2012-texas-droughts.

19. Texas Water [website], "Texas Water Law," Texas A&M University, 2012, accessed August 24, 2012, http://texaswater.tamu.edu/water-law.

20. Jean-Philippe Nicot and Bridget R. Scanlon, "Water Use for Shale-Gas Production in Texas, U.S.," *Environmental Science & Technology* 46, no. 6 (March 20, 2012): 3580–86.

Chapter 4

1. Rachel Carson, *Silent Spring*, 40th anniversary ed. (Boston: Houghton Mifflin, 2002): 234.

2. "Basic Information about Benzene in Drinking Water," USEPA, accessed April 5, 2013, http://is.gd/q4s5pz.

3. Stephen Rassenfoss, "From Bacteria to Barrels: Microbiology Having an Impact on Oil Fields," *Journal*

    of Petroleum Technology 63, no. 11 (November 2011): 32, accessed (pdf) April 5, 2013, http://is.gd/pmU6Sa.

4. Christopher G. Struchtemeyer and Mostafa S. Elshahed, "Bacterial Communities Associated with Hydraulic Fracturing Fluids in Thermogenic Natural Gas Wells in North Central Texas, USA," *FEMS Microbiology Ecology* 81, no. 1 (July 2012): 13–25.

5. Federation of Canadian Municipalities, *Canadian Infrastructure Report Card 2012*, vol. 1, *Municipal Roads and Water Systems* (Ottawa: Federation of Canadian Municipalities, September 2012): 2 (pdf p8), accessed (pdf) April 5, 2013, http://is.gd/IjzaGo.

6. Joe Carroll, "Worst Drought in More than a Century Strikes Texas Oil Boom," *Bloomberg News*, June 13, 2011, accessed August 6, 2012, http://is.gd/1E0EBF; Bruce Finley, "Colorado Farms Planning for Dry Spell Losing Auction Bids for Water to Fracking Projects, *The Denver Post*, April 1, 2012, accessed April 5, 2013, http://is.gd/dNXvCt.

7. Paula Dittrick, "Drought Raising Water Costs, Scarcity Concerns for Shale Plays," *Oil & Gas Journal* 110, no. 7d (July 30, 2012): 20–21, accessed April 5, 2013, http://is.gd/WsSkhd.

Chapter 5

1. ALL Consulting LLC, *The Modern Practices of Hydraulic Fracturing: A Focus on Canadian Resources* (Tulsa, Okla.: ALL Consulting LLC, 2012): 105–107, Table 19, accessed (pdf) April 5, 2013, www.ptac.org/projects/42.

2. David B. McMahon, "How a Gas Well Is Drilled Down into the Ground, and What Can Go Wrong" [captioned slide show] (Charleston, West Va.: West Virginia Surface Owners' Rights Organization, 2007), http://is.gd/ELINtH. See also, e.g., Paul E. Hardisty and Ece Ozdemiroglu, "The Economics of Remediating NAPLs [Non-Aqueous-Phase Liquids] in Fractured Aquifers" (Houston, Tex.: National Ground Water Assn. Conference on Remediation: Site Closure and the Total Cost of Cleanup, November 7–8, 2005): 66–77: pdf at http://is.gd/oR3VMV, citation at http://is.gd/AKfiEM. All three accessed April 5, 2013.

3. James P. Bruce, "Protecting Groundwater: The Invisible But Vital Resource," *Backgrounder* no. 136 (Toronto: C.D. Howe Institute, 2011): 1–2 (pdf 3–4), accessed (pdf) April 5, 2013, www.cdhowe.org/pdf/Backgrounder_136.pdf.

Chapter 6

1. Barbara Crossette, "Severe Water Crisis Ahead for Poorest Nations in Next 2 Decades," *The New York Times*, August 10, 1995, accessed April 5, 2013, http://is.gd/FNHS4W.

2. Ohio Department of Natural Resources, "Senate Bill 315 Information," June 11, 2012, accessed April 10, 2013, http://oilandgas.ohiodnr.gov/laws-regulations/senate-bill-315.

3. Associated Press, "Vermont Fracking Ban: Green Mountain State Is First in U.S. To Restrict Gas Drilling Technique," *Huffington Post*, May 17, 2012, accessed April 5, 2013, http://is.gd/fOoybe.

4. Wade Rawlins (Reuters), "North Carolina Fracking: Governor Beverly Perdue Vetoes Bill," *Huffington Post*, July 1, 2012, accessed April 5, 2013, http://is.gd/8GJnK8.

5. Government of the Northwest Territories, *Northern Voices, Northern Waters: NWT Water Stewardship Strategy* (Yellowknife: Government of the Northwest Territories, 2010): 10–11, accessed (pdf) April 5, 2013, from http://is.gd/Vq0xyH.

6. Reg C. Whiten, "Challenges and Opportunities for Sustainable Water Stewardship in the Upper Kiskatinaw River," *Watermark* 21, no. 1 (BC Water & Waste Assn., Spring 2012): 28–30, accessed (pdf) April 5, 2013, http://is.gd/B3aQVo.

Bibliography

ALL Consulting LLC. *The Modern Practices of Hydraulic Fracturing: A Focus on Canadian Resources*. Tulsa, Okla.: ALL Consulting LLC, 2012. Accessed (pdf) April 5, 2013, www.ptac.org/projects/42.

American Petroleum Institute. "Overview of Industry Guidance / Best Practices on Hydraulic Fracturing." Washington, DC: American Petroleum Institute, 2012. Accesssed (pdf) August 6, 2012, http://is.gd/MpLuSg.

Associated Press. "Vermont Fracking Ban: Green Mountain State Is First in U.S. to Restrict Gas Drilling Technique." *Huffington Post*, May 17, 2012. Accesssed April 5, 2013, http://is.gd/fOoybe.

BC Oil & Gas Commission. FracFocus Chemical Disclosure Registry. Accessed April 5, 2013, http://fracfocus.ca/node/358.

Bott, Robert D. *Evolution of Canada's Oil and Gas Industry*. Calgary: Canadian Centre for Energy Information, 2004.

Bruce, James P. "Protecting Groundwater: The Invisible But Vital Resource." *Backgrounder* no. 136. Toronto: C.D. Howe Institute, February 2011. Accessed (pdf) April 5, 2013, www.cdhowe.org/pdf/Backgrounder_136.pdf.

Canada Water Act, RSC 1985, c. C-11. Accessed March 18, 2013, http://canlii.ca/t/hz6l.

Canada West Foundation. "The Evolution of Water Policy in Alberta." Calgary: Canada West Foundation, 2010. Accessed (pdf) August 6, 2012, http://is.gd/kmMp9m.

———. "Water Pricing: Seizing a Public Policy Dilemma by the Horns." Calgary: Canada West Foundation, 2011. Accessed April 5, 2013, http://cwf.ca/projects/water-pricing.

Canadian Association of Petroleum Producers. "Guiding Principles for Hydraulic Fracturing." Calgary: Canadian Association of Petroleum Producers, 2011. Accessed (pdf) April 5, 2013, via link from http://is.gd/s9LymC.

Canadian Environmental Protection Act, 1999, SC 1999, c. 33. Accessed April 5, 2013, http://canlii.ca/t/51xdb.

Carroll, Joe. "Worst Drought in More Than a Century Strikes Texas Oil Boom." *Bloomberg News*, June 13, 2011. Accessed August 6, 2012, http://is.gd/1E0EBF.

Carson, Rachel. *Silent Spring*. 40th anniversary ed. Boston: Houghton Mifflin, 2002.

Charman, Karen. "Trashing the Planet for Natural Gas: Shale Gas Development Threatens Freshwater Sources, Likely Escalates Climate Destabilization." *Capitalism Nature Socialism* 21, no. 4 (December 2010): 72–82. Accessed (pdf) April 4, 2013, www.karencharman.com/resources/TrashingThePlanet.pdf.

Clean Water Act [U.S.]. 33 U.S.C. §1251 et seq. (1972). Accessed (pdf) April 5, 2013, www.epw.senate.gov/water.pdf. See also variously under USEPA below.

Cooley, Heather, and Kristina Donnelly. *Hydraulic Fracturing and Water Resources: Separating the Frack from the Fiction.* Oakland, Calif.: Pacific Institute, June 2012. Accessed (pdf) April 5, 2013, www.pacinst.org/reports/fracking/full_report.pdf.

Crossette, Barbara. "Severe Water Crisis Ahead for Poorest Nations in Next 2 Decades." *The New York Times*, August 10, 1995. Accessed April 5, 2013, http://is.gd/FNHS4W.

Davis, Charles. "The Politics of 'Fracking': Regulating Natural Gas Drilling Practices in Colorado and Texas." *Review of Policy Research* 29, no. 2 (March 2012): 177–91. Accessed (pdf) April 5, 2013, http://is.gd/12DI0L.

de Loë, Rob. *Toward a Canadian National Water Strategy: Final Report.* Guelph, Ont.: Rob de Loë Consulting Services; Ottawa: Canadian Water Resources Association, 2008. Accessed (cached pdf) April 5, 2013, http://is.gd/QhdDBt.

Dittrick, Paula. "Drought Raising Water Costs, Scarcity Concerns for Shale Plays." *Oil & Gas Journal* 110, no. 7d (July 30, 2012): 20–21. Accessed (paywall) April 5, 2013, http://is.gd/WsSkhd.

Energy Policy Act of 2005 [U.S.]. Pub. L. 109-58, sec. 322. Accessed April 5, 2013, http://is.gd/KUwXXQ.

Energy Resources Conservation Board. "Highlights in Alberta's Energy Development." Accessed April 5, 2013, [archived] http://is.gd/iJdpK2.

———. "Who Is the ERCB and What Is Its Role?" Calgary: ERCB, 2012. Accessed April 5, 2013, http://is.gd/xfvu3Q.

Environment Canada. *Federal Water Policy*. Ottawa: Government of Canada, 1987. Accessed April 5, 2013, http://is.gd/i6bNxS.

———. *Toxic Substances Management Policy*. Ottawa: Government of Canada, 1995. Full text reprint (pdf) accessed April 5, 2013, http://is.gd/YAzMGb.

———. "What Is the Great Lakes Water Quality Agreement?" Ottawa: Government of Canada, 2012. Accessed April 5, 2013, http://is.gd/xSZ90U.

Federation of Canadian Municipalities. *Canadian Infrastructure Report Card 2012*. Vol. 1, *Municipal Roads and Water Systems*. Ottawa: Federation of Canadian Municipalities, September 2012. Accessed (pdf) April 5, 2013, http://is.gd/IjzaGo.

Finley, Bruce. "Colorado Farms Planning for Dry Spell Losing Auction Bids for Water to Fracking Projects." *The Denver Post*, April 1, 2012. Accessed April 5, 2013, http://is.gd/dNXvCt.

Fisheries Act, RSC 1985, c. F-14, ss. 34–42. Accessed April 5, 2013, http://is.gd/VOYra1.

Food & Water Watch. *The Case for a Ban on Gas Fracking*. Washington, DC: Food & Water Watch, 2011. Accessed (pdf) April 5, 2013, via link from www.foodandwaterwatch.org/reports/the-case-for-a-ban-on-gas-fracking.

Gilson, J.C. "Prairie Farm Rehabilitation Administration." Canadian Encyclopedia, n.d. Accessed April 5, 2013, http://is.gd/vYq3cY.

Gobert, Wilf. "The Changing Face of the Canadian Oilpatch." *Canadian Society of Exploration Geophysicists Recorder* 29, no. 3 (March 2004): 46–47. Accessed (pdf) April 5, 2013, via link from http://is.gd/v4KYdE.

Goldstein, Bernard D., Jill Kriesky and Barbara Pavliakova. "Missing from the Table: Role of the Environmental Public Health Community in Governmental Advisory Commissions Related to Marcellus Shale Drilling." *Environmental Health Perspectives* 120, no. 4 (April 2012): 483–86. Accessed April 5, 2013, http://is.gd/KO32Zb.

Gosman, Sara, et al. *Hydraulic Fracturing in the Great Lakes Basin: The State of Play in Michigan and Ohio.* Ann Arbor, Mich.: National Wildlife Federation and University of Michigan Law School, 2012. Accessed (pdf) April 5, 2013, via link from http://is.gd/FHXuZr.

Government of Alberta. "Legislative History of Water Management in Alberta." Accessed April 5, 2013, http://environment.alberta.ca/02265.html.

———. *Water for Life: Alberta's Strategy for Sustainability.* Edmonton: Alberta Environment, 2003. Accessed (pdf) April 5, 2013, via link from http://is.gd/qjiosH.

———. "Watershed Planning and Advisory Councils." Accessed April 5, 2013, www.waterforlife.alberta.ca/01261.html.

Government of the Northwest Territories. *Northern Voices, Northern Waters: NWT Water Stewardship Strategy.* Yellowknife: Government of the Northwest Territories, 2010. Accessed (pdf) April 5, 2013, via link from http://is.gd/Vq0xyH.

Ground Water Protection Council. "About Us." 2012. Accessed April 5, 2013, www.gwpc.org/about-us.

———. "GWPC History." 2012. Accessed April 5, 2013, www.gwpc.org/about-us/gwpc-history.

———. *Modern Shale Gas Development in the United States: A Primer*. Oklahoma City: Ground Water Protection Council, 2009. Accessed (pdf) April 5, 2013, via "Shale Gas Primer" link from www.gwpc.org/resources/publications.

Ground Water Protection Council and Interstate Oil & Gas Compact Commission. "FracFocus Chemical Disclosure Registry: Chemicals & Public Disclosure." Accessed April 5, 2013, http://fracfocus.org/chemical-use/chemicals-public-disclosure.

Hardberger, Amy. "Understanding Water Tradeoffs for Hydraulic Fracturing." *Texas Water Solutions* (blog), September 20, 2012. Environmental Defense Fund. Accessed April 5, 2013, http://is.gd/SByWVV.

Hardisty, Paul E., and Ece Ozdemiroglu. "The Economics of Remediating NAPLs [Non-Aqueous-Phase Liquids] in Fractured Aquifers." Houston, Tex.: National Ground Water Assn. Conference on Remediation: Site Closure and the Total Cost of Cleanup (November 7–8, 2005): 66–77. Accessed (pdf) April 5, 2013, http://is.gd/oR3VMV.

Hinton, Diana Davids, and Roger M. Olien. *Oil in Texas: The Gusher Age, 1895–1945*. Austin: University of Texas Press, 2002.

Hughes, J. David. "Hydrocarbons in North America." In *The Post Carbon Reader: Managing the 21st Century's*

Sustainability Crisis, edited by Richard Heinberg and Daniel Lerch. Healdsburg, Calif.: Watershed Media, 2010. Accessed (pdf) April 5, 2013, www.postcarbon.org/Reader/PCReader-Hughes-Energy.pdf.

Hurlbert, Margot. "Canada's Water Law." Paper prepared for the National Council of Women of Canada, June 2007. Accessed (pdf) April 5, 2013, www.ncwc.ca/pdf/waterlaw.pdf.

International Energy Agency. *Golden Rules for a Golden Age of Gas: World Energy Outlook Special Report on Unconventional Gas*. Paris: International Energy Agency, 2012. Accessed (pdf) April 5, 2013, http://is.gd/aurIWr.

International Joint Commission. "Historical Highlights of the Boundary Waters Treaty." Accessed April 5, 2013, www.ijc.org/en_/IJC_History.

Jaccard, Mark. "Peak Oil and Market Feedbacks: Chicken Little versus Dr. Pangloss." In *Carbon Shift: How the Twin Crises of Oil Depletion and Climate Change Will Define the Future*, edited by Thomas Homer-Dixon with Nick Garrison. Toronto: Random House Canada, 2009.

Jackson, John, and Jane Elder. "Frequently Asked Questions: Great Lakes Water Quality Agreement Renegotiation 2011–2012." Kitchener, Ont., and Buffalo, N.Y.: Great Lakes United, 2011. Accessed (pdf) April 5, 2013, via link from http://is.gd/7ssNqw.

Linux Information Project, The. "The Dismantling of the Standard Oil Trust." 2006. Accessed April 5, 2013, www.linfo.org/standardoil.html.

Loftis, Randy Lee. "EPA: 2 Parker County homes at risk of explosion after gas from 'fracked' well contaminates aquifer." *The Dallas Morning News*, December 9, 2010. Accessed April 5, 2013, [now archived at Homeland1.com] http://is.gd/e9QZRE.

Lustgarten, Abrahm, and ProPublica. "Drill for Natural Gas, Pollute Water." *Scientific American*, November 17, 2008. Accessed April 5, 2013, http://is.gd/IdWxbc.

McMahon, David B. "How a Gas Well Is Drilled Down into the Ground, and What Can Go Wrong" [captioned slide show]. Charleston, West Va.: West Virginia Surface Owners' Rights Organization, 2007. Accessed April 5, 2013, http://is.gd/ELINtH.

Microbes BioSciences. "Enhanced Oil Recovery." Accessed April 5, 2013, http://is.gd/CbjXNL.

Mighty Peace Watershed Alliance. "About MPWA." Accessed April 5, 2013, www.mightypeacewatershedalliance.org/about-mpwa.

Milly, P.C.D. et al. "Stationarity Is Dead: Whither Water Management?" *Science* 319, (February 1, 2008): 573–74. Accessed (pdf) April 5, 2013, http://is.gd/2QUZq1.

Mooney, Chris. "The Truth about Fracking." *Scientific American* 305, November 2011, 80–85.

Morris, T.J., et al. *Changing the Flow: A Blueprint for Federal Action on Freshwater.* The Gordon Water Group of Concerned Scientists and Citizens, 2007. Accessed April 5, 2013, via link from http://poliswaterproject.org/publication/127.

Myers, Tom. "Potential Contaminant Pathways from Hydraulically Fractured Shale to Aquifers." *Ground Water* 50, no. 6 (November/December 2012): 872–82.

New York Marine Sciences Consortium. *An Assessment of Some of the Environmental and Public Health Issues Surrounding Hydraulic Fracturing in New York State*. Stony Brook: New York Marine Sciences Consortium, 2011. Accessed (pdf) April 5, 2013, www.somas.stonybrook.edu/~awp/downloads/NYMSC-FrackingWhitePaper.pdf.

Nicot, Jean-Philippe, and Bridget R. Scanlon. "Water Use for Shale-Gas Production in Texas, U.S." *Environmental Science & Technology* 46, no. 6 (March 2, 2012): 3580–86.

Ohio Department of Natural Resources. "Senate Bill 315 Information." June 11, 2012. Accessed April 5, 2013, http://oilandgas.ohiodnr.gov/laws-regulations/senate-bill-315.

Opalka, Katia. "Oil and Gas: The View from Canada." *Natural Resources & Environment* 26, no. 2 (Fall 2011): 37–40. Reprint (pdf) accessed March 3, 2013, www.cba.org/cba/newsletters-sections/pdf/2012-02-neerls2.pdf.

OPEC. "About Us." Accessed April 5, 2013, www.opec.org/opec_web/en/17.htm.

Patz, Jonathan A., et al. "Climate Change and Waterborne Disease Risk in the Great Lakes Region of the U.S." *American Journal of Preventive Medicine* 35, no. 5 (November 2008): 451–58. Accessed (pdf) April 5, 2013, http://is.gd/zVMJ2w.

Percy, David R. "Responding to Water Scarcity in Western Canada." *Texas Law Review* 83, no. 7 (June 2005): 2091–2107.

Petrutsas, Justin. "Comparing the 2011 and 2012 Texas Droughts." College Station Weather Forecast, *Houston Weather Examiner*, July 26, 2012. Accessed April 5, 2013, www.examiner.com/article/comparing-the-2011-and-2012-texas-droughts.

Phare, Merrell-Ann S.. *Denying the Source: The Crisis of First Nations Water Rights*. Calgary: Rocky Mountain Books, 2009.

Postel, Sandra L., Gretchen C. Daily and Paul R. Ehrlich. "Human Appropriation of Renewable Fresh Water." *Science* (New Series) 271, no. 5250 (February 9, 1996): 785–88. Accessed (pdf) March 3, 2013, http://is.gd/32e23g.

Prairie Provinces Water Board. Master Agreement on Apportionment, 1969, as amended, current to 2009. Scheds. A (Alta./Sask.), B (Sask./Man.). Accessed August 5, 2012, www.ppwb.ca/information/79/index.html.

Probert, Tim. "Shale Gas Fracking: Water Lessons from the US to Europe." *Water & Wastewater International* 27, no. 2 (April/May 2012): 22–24. Accessed April 5, 2013, http://is.gd/4Gw28V.

Rahm, Brian G., and Susan J. Riha. "Toward Strategic Management of Shale Gas Development: Regional, Collective Impacts on Water Resources." *Environmental Science and Policy* 17 (March 2012): 12–23. Accessed April 5, 2013, http://is.gd/grBG2L.

Rahm, Dianne. "Regulating Hydraulic Fracturing in Shale Gas Plays: The Case of Texas." *Energy Policy* 39, no. 5 (May 2011): 2974–81.

Rassenfoss, Stephen. "From Bacteria to Barrels: Microbiology Having an Impact on Oil Fields." *Journal of Petroleum Technology* 63, no. 11 (November 2011): 32. Accessed (pdf) April 5, 2013, http://is.gd/pmU6Sa.

Rawlins, Wade (Reuters). "North Carolina Fracking: Governor Beverly Perdue Vetoes Bill." *Huffington Post*, July 1, 2012. Accessed April 5, 2013, http://is.gd/8GJnK8.

Rister, Carl Coke. *Oil! Titan of the Southwest*. Norman: University of Oklahoma Press, 1949.

Safe Drinking Water Act of 1974 [U.S.]. Pub. L. 93-523, 88 Stat. 1660. Accessed April 5, 2013, http://is.gd/McT20i. See also variously under USEPA below.

Sandford, Robert W., and Merrell-Ann S. Phare. *Ethical Water: Learning To Value What Matters Most*. Calgary: Rocky Mountain Books, 2011.

Shogren, Elizabeth. "EPA Connects 'Fracking' to Water Contamination." *All Things Considered*, National Public Radio, December 8, 2011. Summary, audio and transcript accessed August 6, 2012. http://is.gd/8PlfLY.

Struchtemeyer, Christopher G., and Mostafa S. Elshahed. "Bacterial Communities Associated with Hydraulic Fracturing Fluids in Thermogenic Natural Gas Wells in North Central Texas, USA." *FEMS Microbiology Ecology* 81, no. 1 (July 2012): 13–25.

Texas Water [website]. "Texas Water Law." Texas A&M University, 2012. Accessed August 24, 2012, http://texaswater.tamu.edu/water-law.

Texas Water Development Board. "A Texan's Guide to Water and Water Rights Marketing." Austin: Texas Water Development Board, 2003. Accessed (pdf) April 5, 2013, http://is.gd/6AxraX.

Truhaut, René. "Ecotoxicology: Objectives, Principles and Perspectives." *Ecotoxicology and Environmental Safety* 1, no. 2 (September 1977): 151–73.

USEPA. "Basic Information about Benzene in Drinking Water." Washington, DC: USEPA, 2012. Accessed April 5, 2013, http://is.gd/q4s5pz.

———. "Classes of Wells." Washington, DC: USEPA, 2012. Accessed April 5, 2013, http://water.epa.gov/type/groundwater/uic/wells.cfm.

———. "Great Lakes Water Quality Agreement." Washington, DC: USEPA, 2012. Accessed April 5, 2013, www.epa.gov/glnpo/glwqa.

———. "History of the Clean Water Act." Washington, DC: USEPA, 2012. Accessed April 5, 2013, www.epa.gov/regulations/laws/cwahistory.html.

———. "Hydraulic Fracturing Background Information." Washington, DC: USEPA, 2012. Accessed April 5, 2013, http://is.gd/jwCST4.

———. "Regulation of Hydraulic Fracturing under the Safe Drinking Water Act." Washington, DC: USEPA, 2012. Accessed April 5, 2013, http://is.gd/JAC0d6.

———. "Safe Drinking Water Act, 1974 (SDWA)." Washington, DC: USEPA, 2012. Accessed July 30, 2012, http://is.gd/ri1jXs.

———. "Summary of the Clean Water Act." Washington, DC: USEPA, 2012. Accessed April 5, 2013, www.epa.gov/regulations/laws/cwa.html.

———. "Understanding the Safe Drinking Water Act." Washington, DC: USEPA, 2004. Accessed (pdf) April 5, 2013, under "SDWA Fact Sheets" at http://water.epa.gov/lawsregs/rulesregs/sdwa/index.cfm.

———. "Water: Laws & Executive Orders." Washington, DC: USEPA, 2012. Accessed July 29, 2012, http://water.epa.gov/lawsregs/lawsguidance.

U.S. House of Representatives Committee on Energy and Commerce Minority Staff for Reps. Waxman, Markey and DeGette. "Chemicals Used in Hydraulic Fracturing." April 2011. Accessed (pdf) April 5, 2013, via link from http://is.gd/Hi9Z4W.

Vander Ploeg, Casey. "The Evolution of Water Policy in Alberta." Calgary: Canada West Foundation, 2010. Accessed (pdf) August 6, 2012, http://is.gd/kmMp9m.

Water Act, RSA 2000, c. W-3. Accessed April 5, 2013, http://canlii.ca/t/lg4x.

Whiten, Reg, C. "Challenges and Opportunities for Sustainable Water Stewardship in the Upper Kiskatinaw River." *Watermark* 21, no. 1 (BC Water & Waste Assn.: Spring 2012): 28–30. Accessed (pdf) March 3, 2013, http://is.gd/B3aQVo.

About the author

C. Alexia Lane holds a master's of environmental applied science and management from Ryerson University. She has worked on various water management initiatives in China, Vietnam, Portugal and Canada, and has travelled extensively observing the changes to waterways across the globe.

Other Titles in this Series

The Earth Manifesto

Saving Nature with Engaged Ecology

David Tracey

ISBN 978-1-927330-89-0

The Homeward Wolf

Kevin Van Tighem

ISBN 978-1-927330-83-8

Saving Lake Winnipeg

Robert William Sandford

ISBN 978-1-927330-86-9

Little Black Lies

Corporate and Political Spin in the Global War for Oil

Jeff Gailus

ISBN 978-1-926855-68-4

Digging the City

An Urban Agriculture Manifesto

Rhona McAdam

ISBN 978-1-927330-21-0

Gift Ecology

Reimagining a Sustainable World

Peter Denton

ISBN 978-1-927330-40-1

The Insatiable Bark Beetle

Dr. Reese Halter

ISBN 978-1-926855-67-7

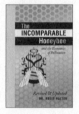

The Incomparable Honeybee

and the Economics of Pollination
Revised & Updated

Dr. Reese Halter

ISBN 978-1-926855-65-3

The Beaver Manifesto

Glynnis Hood

ISBN 978-1-926855-58-5

The Grizzly Manifesto

In Defence of the Great Bear

Jeff Gailus

ISBN 978-1-897522-83-7

Becoming Water

Glaciers in a Warming World

Mike Demuth

ISBN 978-1-926855-72-1

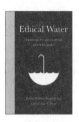

Ethical Water

Learning To Value What Matters Most

Robert William Sandford
& Merrell-Ann S. Phare

ISBN 978-1-926855-70-7

Denying the Source

The Crisis of First Nations Water Rights

Merrell-Ann S. Phare

ISBN 978-1-897522-61-5

The Weekender Effect

Hyperdevelopment in Mountain Towns

Robert William Sandford

ISBN 978-1-897522-10-3

RMB saved the following resources by printing the pages of this book on chlorine-free paper made with 100% post-consumer waste:

Trees · 7, fully grown
Water · 3,107 gallons
Energy · 3 million BTUs
Solid Waste · 208 pounds
Greenhouse Gases · 573 pounds

CALCULATIONS BASED ON RESEARCH BY ENVIRONMENTAL DEFENSE AND THE PAPER TASK FORCE. MANUFACTURED AT FRIESENS CORPORATION.